Estates Gazette

PLANNING LAW REPORTS

Estates Gazette

Planning Law Reports

1989
Volume 2

Edited by
Barry Denyer-Green LLM PhD FRICS
of the Middle Temple, barrister

Ernest Speller
Publisher's Editor

Routledge
Taylor & Francis Group
New York London

EG Books is an imprint of Routledge

2 Park Square, Milton Park, Abingdon, Oxon OX14 4RN
711 Third Avenue, New York, NY 10017, USA
Routledge is an imprint of Taylor & Francis Group, an Informa business.

First published 1989
ISBN for complete set of 3 volumes: 0 7282 0142 9
ISBN for this volume: 0 7282 0139 9

ISSN 0954-1977

Transferred to digital printing 2007

PLANNING LAW REPORTS

[1989] 2 PLR

CASES REPORTED IN VOLUME 2

INDEX OF SUBJECT-MATTER

CASES JUDICIALLY CONSIDERED

For case references see pages shown in bold type PAGES

Smith *v* Oliver

QUEEN'S BENCH DIVISIONAL COURT
Farquharson and Tudor Price JJ
April 25 1985

Trees — Conservation area — Trees topped without consent — Prosecution — Defence that trees "have become dangerous" — Adjoining public highway — Trees not themselves dangerous — Trees presenting danger by reason of position and size — Whether within statutory defence — Whether "have become dangerous"

A

The respondent was the owner of a suburban garden in Chertsey, Surrey. His
property was within a designated conservation area the consequence of which
is that it is an offence to carry out certain operations to trees without the
consent of the local planning authority: see sections 60 and 61A(1) of the
Town and Country Planning Act 1971. The respondent topped or lopped
several trees without consent and informations were laid against him alleging
an offence under section 61A(1). His defence was that by reason of section B
60(6) a person cannot be guilty of an offence where he does any of the
prohibited operations to trees that "have become dangerous", and that such
trees may be so described where they are dangerous by reason of their position
or size in relation to an adjoining public footpath. The justices found that the
trees had become dangerous and dismissed the informations; the prosecutor
appealed by way of a case stated.

Held The appeal was dismissed. C
Whether trees have become dangerous for the purposes of the statutory
exception is a question of fact for the justices: see p 3E. They should take the
sensible approach of a prudent citizen in deciding whether trees are
dangerous; there must be a present danger and it is not limited to disease or
damage to the trees themselves. The threatened danger does not actually have
had to have occurred; it is sufficient to find that, by virtue of the state of the
trees, their size, their position and such effect as any of those factors have, one D
can properly conclude that the trees have become dangerous. The justices
should look at what is likely to happen, such as injury to a passing pedestrian:
see pp 3E-4A and 4H.

No cases are referred to in the judgments

Appeal by way of a case stated E
This was an appeal by way of a case stated from the decision of the Chertsey Justices dismissing five informations laid against the respondent that he had topped certain trees in a conservation area contrary to section 61A(1) of the Town and Country Planning Act 1971.

Brian Ash (instructed by the solicitor to Runnymede Borough Council) appeared for the appellant.
Norman Wise and Nigel Pleming (instructed by Engall & Freeman, of Egham, Surrey) appeared for the respondent. F

The following judgments were delivered.

FARQUHARSON J: This is an appeal by way of case stated from a decision of the Chertsey Justices dismissing five informations laid against the respondent. Each of those informations alleged against him that he had topped trees in a conservation area contrary to section 61A(1) of the Town and Country

Planning Act 1971. The trees which were the subject of the informations were all situated or growing in a small suburban garden in Chertsey where the respondent lived.

Turning to the statutes, the section under which these offences were laid, section 61A(1), reads as follows:

> Subject to the provisions of this section, any person who, in relation to a tree to which this section applies, does any act which might by virtue of section 60(1)(*a*) above be prohibited by a tree preservation order shall be guilty of an offence.

Subsection (2) of that same section spells out to what the section does apply and says that it "applies to any tree in a conservation area but in respect of which no tree preservation order is for the time being in force".

Section 60 of the Town and Country Planning Act 1971 gives the local planning authority power to make tree preservation orders where it is expedient to do so. The effect of section 61A just recited is to extend that protection to conservation areas where the trees are not the subject of specific orders under section 60. There is a statutory instrument* which specifically enables anyone charged with this offence to rely on the defences which are provided in section 60, and the particular subsection we are concerned with is subsection (6), which is in the following terms:

> Without prejudice to any other exemptions for which provision may be made by a tree preservation order, no such order shall apply to the cutting down, uprooting, topping or lopping of trees which are dying or dead or have become dangerous, or the cutting down, uprooting, topping or lopping of any trees in compliance with any obligations imposed by or under an Act of Parliament or so far as may be necessary for the prevention or abatement of a nuisance.

The way the case was run before the justices was that the respondent had no right to cut down the trees as he had no licence or permission from the local planning authority so to do; but he contended that he was justified in lopping the trees in his garden because he was protected by the subsection: he contended that the trees had become dangerous within the meaning of that subsection.

I now turn to the facts, though very few have been recited. It appears that the trees concerned were a sycamore, three hornbeams and a Norway maple, all of which were in the garden of the respondent near to the front fence, on the other side of which was the public highway bordered by a concrete pavement. The Norway maple tree seems to have faded almost into extinction during the course of the proceedings. No findings were made about it and very little reference was made to that tree in the course of this appeal. Dealing with that just for the moment, one can only say that here is no evidence upon which we can properly make any order with regard to the Norway maple; but the important issue arises as a result of the treatment of the other trees.

The justices dealing with the matter, besides hearing evidence about the condition of the trees in court, also took the trouble to attend the site and to inspect the now lopped trees themselves. It appears that on October 28 1983, the respondent cut the trees to a height of 13, 14 or 15 ft respectively; clearly he is in breach of section 61A if he is not entitled to rely upon the exemption.

The only findings of fact which will assist us in our present consideration of the appeal are contained in para 2(*d*) and (*e*) of the justices' case:

> The sycamore and three hornbeams were very close to a wooden fence, in two cases touching it. Before they were topped, a strong wind would cause them

*Town and Country Planning (Tree Preservation Order) Regulations 1969 (SI 17).

> to sway against the fence. The four trees next to the fence, ie the sycamore and three hornbeams, were likely to push it over as they grew, one having already broken the arris rail of the fence, and the roots were likely to push up the pavement on the other side of the fence.

The justices, having set out those findings, then proceeded to recite all the evidence, or at least a large part of it, which had been given by the witnesses. There is no indication in the case as to what part of that evidence was accepted, save in so far as we can rely upon the contentions of the parties and the justices' own conclusions. They said, and it is in rather wise terms, that "Having considered all the evidence and having gone to view the trees in question we were of the opinion that the trees were dangerous at the time the respondent had them cut therefore he had made out the defence available under the Regulations. Accordingly we dismissed all five informations . . . ".

It has been said time and again that where justices have heard all the evidence their duty is not to recite that evidence when preparing the case stated, but to set out their findings of fact. The only time when it is necessary to recite evidence is when it is submitted that the finding of fact cannot be justified on the evidence available. The justices should only have set out here the facts upon which they relied to justify their opinion that the defence provided by the exemption to section 60(6) had been made out.

This is not a case where it is desirable to involve or indeed incur further expense, so we shall have to do the best we can with the material that is available to us.

I now turn back to the relevant subsection, that is to say subsection (6) of section 60 of the Town and Country Planning Act 1971. The phrase at the very beginning of that subsection removes the prohibition made by section 60 and section 61A in relation to cutting down, topping or lopping of trees "which are dying or dead or have become dangerous". In my judgment, that must be a question of fact for the justices. The approach which they should make is the everyday sensible approach of a prudent citizen looking at the trees in question and deciding in his own mind whether he can properly say those trees are dangerous. The existence of the danger must be a present danger. He must be able to say that the existing condition is one of danger in relation to the tree or trees which he is examining. Of course that does not mean that the danger which has been threatened has actually occurred. It is not necessary to show that the tree has fallen or that its roots have disturbed the foundation of the house, fence or the pavement nearby. The justices must be in a position to say to themselves that "having regard to the state of the tree, its size, its position and such effect as any of those factors have so far had, we can properly come to the conclusion that the tree has now become dangerous". It may be that the condition of the tree itself is perfectly healthy, but because of those other factors the magistrates may be able none the less to say that the tree is in the condition described by the subsection, that is to say it has become dangerous.

In deciding that question, the magistrates are entitled to look at what is likely to happen. If the tree has already shown signs of disturbing a fence or a pavement or indeed the house itself, it does not need the justices or indeed anyone concerned with the treatment of the tree to wait for those events actually to occur, namely for the fence to fall down on some passing pedestrian or the condition of the pavement to be such that somebody falls and is injured or that the house begins to subside. The present danger is one that is constituted by the conditions at the time the tree is being examined, and the fact that these other matters may shortly arise is something that can properly be taken into account when the justices come to their decision.

Of course it does not mean that any individual who owns trees and wishes to remove them can say to himself, "having regard to their position, the way they have been planted and their unsuitability, I can see that in a number of years they are going to constitute a danger'' and then cut them down. One has to look at the position at the time. If such damage is far off, remote and not immediate in the sense that I have just described, the trees do not come within the meaning of the exemption.

Having regard to the correct approach as we find it to be, were the justices justified in coming to the conclusion they did in this case? For the reasons I have explained and the manner in which the case is prepared, this has caused us more anxiety than perhaps any other aspect. Bearing in mind the matters that I have referred to and the features of the evidence which the justices plainly must have accepted to have come to the conclusion that they did, for my part I conclude that there was evidence that those trees had become dangerous. I bear very much in mind the fact that the justices took the trouble to go and look at the trees themselves, and sensible people would be able to conclude by their own observations very quickly whether the trees were properly so described.

It is the fact, as appears from the evidence, that the way the respondent dealt with these trees, cutting them down to the level he did, did not improve the position at all — an expert says that. The condition of the trees was such and the height at which he left them was such that the matters which were causing him anxiety remained.

Mr Ash, for the appellants, submitted to us that one should look at that feature in deciding whether the terms of the exempting clause have been established by the respondent. He said it is not much good, even if a danger were present, if the owner of the trees can then mutilate the trees in a way that has no beneficial effect from the point of view of reducing the danger.

That is an attractive argument. But for my part it falls down when one looks at the terms of the subsection itself, because there is no reference in the phrase ''trees . . . have become dangerous'' to the suggestion made by counsel that that entitles the owner or the person concerned with the treatment of the trees only to reduce them to the extent that the danger is removed. One is fortified in this construction by looking at the further words in the subsection, because when the subsection deals with the question of the abatement of a nuisance, it uses these words: ''or the cutting down, uprooting, topping or lopping of any trees . . . so far as may be necessary for the prevention or abatement of a nuisance''. So the words ''so far as may be necessary'' are introduced by the draftsman when dealing with the question of nuisance, but they are not included when referring to the trees being dangerous. I for my part feel bound to reject that submission.

In all those circumstances and for the reasons that I have given, I have come to the conclusion and propose that this appeal should be dismissed.

TUDOR PRICE J: I agree with everything which has fallen from my Lord. I would wish to add only a few words, because it was at one stage argued on behalf of the appellant that the words ''have become dangerous'' in section 60(6) of the 1971 Act should be construed in a very restricted manner.

In my judgment, that contention was wrong. A tree has become dangerous at the moment when it constitutes a present danger to persons or property. That may be not only as a result of disease or damage, it may be by reason of its size and location. For example, a tree planted at a roadside may by its very size be judged to have become dangerous in the sense that in a high wind it, or part of it, may fall across the road to the danger of traffic. A common case is that of a forest tree planted in a suburban garden which begins to dominate and tower above the house roof so that in high wind it might fall and hole the roof or its roots begin to undermine the foundation. In each case it is a question of fact

for the justices to see whether or not the person who has felled or lopped A
without the necessary permission is able to bring himself within the exemption of the Act.

Appeal dismissed.

Garvey *v* Clydebank District Council

COURT OF SESSION: INNER HOUSE
Lord Ross (Lord Justice Clerk), Lords Brand and McDonald
October 4 1988

Lands Tribunal — Claim for compensation — Statutory time-bar on references to the tribunal — Reference out-of-time — Whether tribunal in error in refusing to determine reference

The respondent district council acquired land from the appellant under a general vesting declaration of February 1977. Negotiations with the district valuer proceeded for several years. In October 1986 the appellant referred the disputed claim for compensation to the Lands Tribunal for Scotland, and the tribunal refused to accept the application on the ground that the claim was time-barred. The appellant appealed by way of a special case.

Held The appeal was dismissed.

The six-year period, within which a question of disputed compensation arising out of an acquisition of an interest in land can be referred to the tribunal, runs from the date at which the claimant first knew, or could reasonably be expected to have known, of the vesting of his interest: para 36, Schedule 24 to the Town and Country Planning (Scotland) Act 1972. It followed that as the latest date the claimant could have known of the vesting was February 1977, the date of confirmation of the vesting declaration, his reference should have been made before February 1983. The tribunal was entitled to refuse jurisdiction in respect of a reference made in October1986.

Case referred to in the opinion
Apostolic Church Trustees v *Glasgow District Council* 1977 SLT (LT) 24

Appeal by way of special case from the Lands Tribunal for Scotland
This was an appeal by way of a special case from the Lands Tribunal, who had decided that the appellant's application to refer a question of disputed compensation was time-barred.

The appellant appeared in person.
James Peoples (instructed by Stathern & Blair WS) appeared for the respondents.

The following opinion of the court was delivered.

LORD JUSTICE CLERK (LORD ROSS): This is a special case stated by the Lands Tribunal for Scotland under section 13 of the Tribunals and Inquiries Act 1971 for the opinion of the Court of Session at the request of the appellant, who is Laurence Garvey, against the respondents, who are Clydebank District Council.

Although the appellant, who appeared in person, sought to raise a large number of matters in his address to the court, the sole issue raised in the special case is whether the Lands Tribunal for Scotland were correct in law in holding that the appellant's application to the tribunal was time-barred and that his application should be dismissed.

The background to this appeal is fully set forth in the statement of facts contained in the special case. The first matter which the appellant raised was that a plea-in-law raising the issue of time-bar had been added by way of amendment, and he contended that the respondents ought never to have been allowed to amend their pleadings by adding this plea-in-law. It is plain from the terms of the case that the tribunal heard parties before determining to allow the amendment. Whether or not to allow the pleadings to be amended in this way was a matter for the discretion of the tribunal, and nothing which the appellant has said would persuade us that we would be justified in interfering with the tribunal's decision in this respect. It should be added that there is no question in the case raising the issue of whether the amendment should have been allowed. A

It was obvious that the appellant feels a strong sense of grievance. He maintained that negotiations had proceeded between himself and the district valuer over some 11 years and that a neighbouring owner whose subjects were less extensive than the appellant's subjects had received over £25,000 in compensation whereas the largest sum which the district valuer had offered the appellant was £4,000. He thus contended that he had been dealt with unjustly. Again it must be pointed out that this is not an issue which is raised before us and that the sole issue before this court is the matter of time-bar. B C

The provisions relating to time-bar are to be found in para 36 of Schedule 24 to the Town and Country Planning (Scotland) Act 1972. That paragraph is in the following terms:

> (1) The time within which a question of disputed compensation, arising out of an acquisition of an interest in land in respect of which a notice to treat is deemed to have been served by virtue of paragraphs 6 to 8 above, may be referred to the Lands Tribunal shall be six years from the date at which the person claiming compensation, or a person from whom he derives title, first knew, or could reasonably be expected to have known, of the vesting of the interest by virtue of those paragraphs. D
>
> (2) In reckoning the period of six years referred to in sub-paragraph (1) of this paragraph, no account should be taken of any period during which the person claiming compensation or the person from whom he derives title was in minority or less age or was under legal disability. E

In the light of the findings in this case the tribunal held that the appellant was aware of the vesting of his interest from 1974 or in any event from the date of the decision of the Second Division of the Court of Session on February 8 1977 (or February 19 1977) refusing his appeal against the granting of decree against him in an action of summary ejection. The tribunal then stated: F

> Accepting that the latest date when he could be considered to have known about the confirmation of the vesting was 18 February 1977, his time limit in terms of paragraph 36 of Schedule 24 was 18 February 1983. As his application to the Tribunal was not formally made until 17 October 1986, although there had been correspondence with the clerk to the Lands Tribunal for Scotland since 2 October 1985, the application was well outside the time limit for doing so. G

We are quite satisfied that on the facts the tribunal were fully justified in arriving at this conclusion. The appellant maintained that the six-year period did not apply in the way that the tribunal had applied the period, and he contended that the six-year period should run from the date when negotiations were broken off. His submission was that the date when he became aware of the general vesting declaration did not enter into the matter at all. He claimed to find support for his submission from *Apostolic Church Trustees* v *Glasgow District Council* 1977 SLT (LT) 24. In particular, he sought to rely on a passage from the opinion of the tribunal at p 26: H

> With regard to the various anomalies raised by claimant's counsel to support a starting point for the time bar as at the date of dispossession, we generally agree with the acquiring authority's contention that it would be technically competent to make a reference to the Tribunal prior to dispossession — and

> A With regard to the various anomalies raised by claimant's counsel to support a starting point for the time bar as at the date of dispossession, we generally agree with the acquiring authority's contention that it would be technically competent to make a reference to the Tribunal prior to dispossession — and therefore prior to expiry of the six-year time limit. Nevertheless one would not expect this to be done, even by a professional adviser, if negotiations were proceeding normally.

B As we understood it, the appellant sought to take from that passage that there could be no question of time-bar running so long as negotiations were continuing. In our opinion, however, he has misunderstood this passage. In a later passage in the opinion in that case it is stated:

> Whatever may be the alleged compensation anomalies resulting from recent United Kingdom planning legislation which introduced the new expedited
> C procedure, the Tribunal have to apply the actual wording of para 6(1) of the Sixth Schedule as they find it. The Schedule plainly states that the six-year period is to run not from when the local authority entered upon and took possession of the land but from the date at which the claimant first knew, or could reasonably be expected to have known, of the vesting of his interest under para 3 of the Schedule.

That case concerned the Town and Country Planning (Scotland) Act 1945,
D Schedule 6, para 6(1), but these provisions are substantially the same as the provisions of para 36 of Schedule 24 to the Act of 1972. We entirely agree with what the tribunal said in that case regarding their obligation to apply the actual wording of the Schedule as they found it. So, applying the words of the Schedule in this case, we are entirely satisfied that the tribunal were correct when they concluded that the six-year time period had elapsed before the appellant made his application to the tribunal on October 17 1986. It follows
E that the question in the case falls to be answered in the negative and the appeal refused.

Appeal refused.

Steinberg and another *v* Secretary of State for the Environment and another

QUEEN'S BENCH DIVISION
Mr Lionel Read QC (sitting as a deputy judge of the Queen's Bench Division)
November 25 1988

Conservation area — Application for planning permission — Appeal to Secretary of State — Planning permission granted on appeal — Challenge to decision by neighbourhood association — Whether inspector failed to accord reasonable time for representations — Whether inspector failed to take into account desirability of preserving or enhancing character of conservation area A

The Secretary of State for the Environment, by his inspector, allowed an appeal against a refusal of planning permission by Camden London Borough Council, and granted planning permission by a decision letter of April 21 1988 for a two-storey house on a vacant site in Kentish Town. The appeal site lies in a designated conservation area.

The Leighton Road Neighbourhood Association, of which the applicants were members, made representations to the local planning authority, but were not informed of the appeal until shortly before the inspector was due to make his site visit. B

The applicants appealed under section 245 of the Town and Country Planning Act 1971 contending that: (1) the inspector failed to accord them reasonable time to make representations prior to his site visit, and as a result they have been prejudiced; and (2) the inspector failed to take into account, or give proper weight to, the duty imposed by section 277(8) of the 1971 Act to pay special attention to the desirability of preserving or enhancing the character and appearance of the conservation area. C

Held 1. The appeal was allowed on the second of the two grounds. Nowhere in his decision letter did the inspector mention section 277(8) of the 1971 Act or his duty thereunder to pay "special attention . . . to the desirability of preserving or enhancing its character or appearance in the exercise D with respect to any buildings or other land in that area of any powers under this Act . . .". Although that omission did not determine the issue, as the obligation of the inspector was not merely to refer to the duty but to discharge it. If the inspector makes no reference to the duty, it must be apparent from the decision that it has been discharged or otherwise there will be an error of law, either because the necessary inference must be made that he has not discharged it, or because he has failed to give adequate reasons for a decision which he E could not validly reach without discharging that duty: see pp 12E-13A.

It was apparent from his decision letter that the inspector had considered the question of harm to the conservation area, but not to the desirability of preserving or enhancing the character or appearance of the conservation area. It could not be inferred that the inspector had discharged the duty imposed by section 277(8) of the 1971 Act.

2. Although there was a breach of reg 5 of the Town and Country Planning F (Appeals) (Written Representations Procedure) Regulations 1987 in failing to give notice to the applicants, the applicants had suffered no prejudice as a result, and the appeal would not have been allowed on this ground: see p 13D.

No cases are referred to in the judgment

Appeal under section 245 of the Town and Country Planning Act 1971
This was an appeal under section 245 of the Town and Country Planning Act 1971

A by two members of the Leighton Road Neighbourhood Association against a
decision of the Secretary of State for the Environment, by his inspector, whereby
he allowed an appeal against a refusal of planning permission by the second
respondents, Camden London Borough Council.

Professor Hannah Steinberg, appeared in person on behalf of both applicants, and was advised by Pothecary & Barratt.
B *Robert Jay (instructed by the Treasury Solicitor) appeared for the first respondent, the Secretary of State for the Environment.*
The second respondent, Camden London Borough Council, did not appear and were not represented.

The following judgment was delivered.

MR LIONEL READ QC: This is an application by Hannah Ruth Steinberg
C and Elizabeth Ann Bowen Sykes under section 245 of the Town and Country
Planning Act 1971 for an order to quash a decision by one of the inspectors of
the Secretary of State for the Environment, the first respondent, given by
letter dated April 21 1988 allowing an appeal by Devon & Wood Property Ltd
against a refusal of planning permission by Camden London Borough
Council, and granting such permission.

D Devon & Wood Property Ltd applied for planning permission in February
1987 to erect a two-storey house on land referred to as 18 Torriano Cottages in
Kentish Town in the London Borough of Camden (which I shall call "the
appeal site"). This is an essentially unused, derelict and overgrown small piece
of land which lies at the rear of 122 Leighton Road and is separated from the
garden of that property by a fence. Torriano Cottages run, at this point,
parallel to and south of Leighton Road. Both roads have houses fronting
E them, and the gardens of the two rows of houses essentially back on to each
other. The appeal site lies between 15 and 19 Torriano Cottages. The first
applicant, Prof Steinberg, lives at no 19.

The appeal site lies in a recently designated small conservation area, the
boundaries of which are drawn tightly around Torriano Cottages and the
short section of Leighton Road, the houses on which back on to Torriano
Cottages. The council refused planning permission for the development on
F October 27 1987, and the developers appealed. The appeal was dealt with by
written representations.

An association, of which both applicants are members, called the Leighton
Road Neighbourhood Association, had made representations to the council
against the proposed development. They were not, however, informed of the
appeal until well after it was made and very shortly before the inspector was
due to make his site visit. They none the less made haste to bring together
G written representations for the inspector to consider, but he did not have them
before his site visit. He did, however, have them and did consider them before
he wrote his decision letter in April 1988.

By that letter he allowed the appeal and granted permission. The applicants
apply to this court to quash that decision as persons aggrieved. The Secretary
of State accepts their *locus* to move the court.

H The applicants appear in person and Prof Steinberg has presented their
case. I hope that I am not taken to be condescending when I say that her
presentation of their case, both in writing beforehand and orally to me, has
been impeccable. There need be no fear that the merits of the points made have
suffered for lack of legal representation.

The applicants' skeleton argument, to which Prof Steinberg spoke, makes it
clear that by their notice of motion they take two points.

1. The inspector failed to accord them a reasonable time, or any time, to

make representations prior to his site visit and as a result they have been A
prejudiced.

2. The inspector failed to take into account, or give proper weight to, the duty imposed by section 277(8) of the 1971 Act to pay special attention to the desirability of preserving or enhancing the character and appearance of the conservation area.

The second ground raises a substantive, as opposed to a procedural, point, and I deal with it first. It is a short point. Section 277(8) provides as follows: B
"Where any area is for the time being designated as a conservation area, special attention shall be paid" — and I pause to stress the word "shall" — "to the desirability of preserving or enhancing its character or appearance in the exercise with respect to any buildings or other land in that area, of any powers under this Act . . .". More than that I need not read. Other enactments not relevant to this matter are referred to. C

It is not in doubt that in determining the appeal by the developers and granting planning permission, the inspector was exercising powers under the 1971 Act, such that it was his duty to pay special attention to the matters specified by section 277(8).

Nowhere in his decision letter, however, does he mention this subsection or his duty thereunder, either in terms or by the use of any language from which it might, in my judgment, reasonably be inferred that he was intending to refer D
to it. That omission nevertheless does not determine the issue raised by the applicants.

The obligation imposed by the statute is to pay special attention to the desirability of preserving or enhancing the character of the conservation area in exercising the power to determine the developers' appeal. For the due discharge of that duty the inspector did not, in my judgment, need to say that he was discharging or was conscious of that duty. What matters is that he does E
discharge that duty, and if he makes no reference to it in his decision it must, in my judgment, be apparent from his decision that he has discharged it or otherwise there will be error in law: either because the necessary inference must be that he has not discharged it or because he has failed to give adequate reasons for a decision which he could not validly reach without discharging that duty. F

Passages in the decision letter relevant to this issue are to be found in paras 6, 11 and 12, which I read:

> 6. From my observations of the site and its surroundings and from the representations received I consider that the main issues to be decided in this appeal are, firstly, whether the proposal would constitute overdevelopment of the site and be unneighbourly in the area, and secondly, whether the proposed development would harm the character of the conservation area. G

He dealt then with the first of those issues, which is of no relevance to my task, and at para 11 turned to the second issue:

> Turning to the second main issue I found the Cottages to be a close-knit urban community in a quiet backwater, comprising a mixture of old and new houses. Whilst the unmade-up nature of the road and the abundance of trees and shrubs impart a rustic charm to this part of the conservation area, I consider that the appeal site, which consists of derelict, wrecked garages and H
> weed-infested areas, does not add to the character of the area. In my view it detracts considerably from both the residential amenity and the visual character of the locality. It appears to me that all of the modern houses on the north and east sides of Torriano Cottages have been built on the back gardens of those houses facing on to Leighton Road and Torriano Avenue and that the appeal site and the plot east of no 19 are the only remaining undeveloped plots. The latter occupies a prominent corner position and contains tall trees

> and shrubs, having been incorporated into the curtilage of no 19, whilst the appeal site contributes little to the surroundings. I consider the design of the proposed 2-storey house to be an acceptable infill between the 3-storey and one-and-a-half-storey buildings. I do not regard the elevational treatment as exceptional but the scale and bulk of the proposed house would provide an acceptable visual link between the unusual single-storey patio design of no 15 and the prominently sited 3-storey no 19.
>
> 12. I have considered the possibility of imposing conditions upon the proposal as suggested by the Council and I intend to impose conditions on the choice of facing materials to ensure that these are sympathetic in the conservation area; on hard and soft landscaping to ensure the provision of the car parking space and to screen the rear garden from the adjoining properties; and to remove permitted development rights to control future privacy and amenity in the locality.

Mr Jay, for the Secretary of State, submits that it is apparent from these paragraphs that I have read that the inspector did have his statutory duty in mind. The second issue defined in para 5 was, as Mr Jay put it, "shorthand form" for the statutory duty. Moreover, while this issue is on its face couched in terms of "harm", it is plain that by the time he reaches para 11 he is looking at preservation as required by the statute.

Mr Jay draws attention to following parts of that paragraph in particular. First, ". . . does not add to the character of the area". Second, ". . . the appeal site contributes little to the surroundings . . .". Third, "I consider the design of the proposed 2-storey house to be an acceptable infill between the 3-storey and one-and-a-half-storey buildings".

Drawing attention also to the reference in para 12 to the need to ensure that the facing materials are sympathetic in the conservation area, Mr Jay submits that there is an irresistible inference that he had regard to his duty under section 277(8).

He further submits that there was no duty on the inspector to pay attention to any question of enhancing the character of the conservation area, since he had paid attention to preserving its character. If, however, it was necessary for the inspector to have done so, Mr Jay submits that he did do so and points to his finding in para 11 on the derelict, weed-infested state of the site with wrecked garages.

I do not accept Mr Jay's submissions. There is, in my judgment, a world of difference between the issue which the inspector defined for himself — whether the proposed development would "harm" the character of the conservation area — and the need to pay special attention to the desirability of preserving or enhancing the character or appearance of the conservation area. In short, harm is one thing; preservation or enhancement is another. No doubt the inspector has demonstrated his concern that the character of the conservation area should not be harmed. That, in my judgment, is not the same as paying special attention to the desirability of preserving or enhancing that character as well as its appearance. The concept of avoiding harm is essentially negative. The underlying purpose of section 277(8) seems to me to be essentially positive.

Against an explicit background where the identified issue is whether the character of the conservation area would be harmed, I do not agree that the inference that the inspector has paid special attention to the desirability of preserving or enhancing the character or appearance of the conservation area — the four remaining elements in the duty — can be drawn from his definition of that issue or from the manner in which he addresses it at para 11, let alone that it is irresistible.

In my judgment, therefore, the inspector is shown to have misdirected himself on a point of law. It is not for the court to conjecture whether he would or might have reached the same conclusion if he had directed himself properly or, therefore, whether he, or some other inspector appointed by the Secretary of State to redetermine the application, might or might not reach the same conclusion when the court remits the matter. That is a matter for the inspector's judgment. But where the statute lays a clear and positive duty as to the manner in which, *inter alia*, the power of determining applications for planning permission for development in a conservation area are to be fulfilled, the court's concern is to ensure that that duty is complied with. The consequences of compliance are for the authority exercising the power in discharge of their statutory duty.

The question whether this statutory duty can be discharged by considering only whether development would preserve the character of a conservation area, and granting permission if it would, does not, therefore, in the event fall for decision. Since, however, the point has been raised and the applicants are concerned to have it decided, I am bound to say that I, for my part, would find that a surprising result of the subsection. On that basis the desirability of enhancing a conservation area would rarely fall to be addressed on an application for planning permission. The question is, however, one of complexity and importance and best decided after full argument when it is directly in point.

I turn to the first ground. The Secretary of State does not dispute that there was a breach by the council of regulation 5 of the Town and Country Planning (Appeals) (Written Representations Procedure) Regulations 1987 in failing to give notice to the applicants in due time of the developers' appeal. I am, however, not satisfied that the applicants suffered prejudice thereby. They made, through their association, a very full written representation which the inspector received well before he wrote his letter and took into account.

Prof Steinberg says that he was wrong, as a matter of fact, in concluding in para 11 that the appeal site and the plot east of no 19 are the only remaining undeveloped plots. She points on the plan of the conservation area to what appears to be an undeveloped plot by Davies Cottages, just round the corner from the appeal site fronting Torriano Cottages. But I cannot decide on the evidence before the court whether he was wrong as alleged. Moreover, it seems clear from his description of the area at para 2 of his letter that he was aware that the whole length of Torriano Cottages, which he plainly saw, was in the conservation area.

In the result, on the second ground of the notice of motion, I allow the application and quash the decision of the inspection.

Application allowed with costs.

Sweet *v* Secretary of State for the Environment and another

QUEEN'S BENCH DIVISION
Schiemann J
January 24 1989

Wildlife and Countryside Act 1981 — Site of special scientific interest — Conservation Order — Operations specified as potentially damaging — Whether order can include land not of national importance — Whether
A **operations includes agricultural activities**

The applicant is the owner of three meadows and adjoining ditches within Westhay Moor, Somerset. In 1985 the applicant was informed by the second respondents, the Nature Conservancy Council, that in their opinion Westhay Moor was of special scientific interest as it "forms part of the nationally important grazing marsh and ditch systems of the Somerset Levels and Moors . . . [and] supports a nationally outstanding community of terrestrial and
B aquatic vertebrates". In June 1987 the Secretary of State for the Environment, in accepting the recommendations of an inspector who had held a local inquiry, made an order under section 29 of the Wildlife and Countryside Act 1981 in respect of Westhay; the order specified certain operations which, by virtue of section 29(3) of the 1981 Act, may not be carried out otherwise than permitted by the Act.

The applicant challenged the validity of the order, contending that: (1) land
C which was not of national importance could not be included within the order with other land that was of such importance, and there were other powers available to the Nature Conservancy Council to protect land not of national importance; and (2) the order specified a number of operations which, under the terms of section 29(3) of the Act, could not be prohibited operations as they were agricultural activities that would not result in some physical alteration to the land of some permanence. The Act made a distinction between "activities" and "operations", and "operations" had a particular
D meaning in planning legislation that did not include agricultural activities.

Held The application was dismissed.

1. On the evidence before him, the inspector was entitled to conclude that while only part of the order land was of national importance, the whole of the land constituted a single environment. It followed that the Secretary of State was entitled to draw the boundaries of the order land where he did and to treat
E the site as a whole and find that it was of national importance: see p 21A-B.

2. Where an applicant fails to put before an inspector at an appeal a ground of objection, there must be hesitation about granting relief on judicial review. Although the word "operations" is used in the 1981 Act, and is a term of art in planning legislation, such legislation is not of any assistance in construing the 1981 Act. Looking at the purpose of the Act, the phrase in section 29(3), "any operation which . . . appears to the Secretary of State to be likely to destroy . . . flora" etc, is wide enough to cover the matters specified in the schedule to
F the order as prohibited operations: see p 21E-G.

Case referred to in the judgment

Parkes v *Secretary of State for the Environment* [1978] 1 WLR 1308; [1979] 1 All ER 211; (1978) 77 LGR 39; 36 P&CR 387; [1978] EGD 973; 248 EG 595; [1979] JPL 33, CA

Application under Schedule 11 to the Wildlife and Countryside Act 1981

This was an application under Schedule 11 to the Wildlife and Countryside Act

1981 to challenge the decision of the first respondent, the Secretary of State for the A
Environment, to make a conservation order under section 29 of the Act following notification by the second respondents, the Nature Conservancy Council.

John Howell (instructed by Turner Kenneth Brown, for Whitehouse & Hawkes, of Highbridge, Somerset) appeared for the applicant.

Christopher Katkowski (instructed by the Treasury Solicitor) appeared for the first respondent, the Secretary of State for the Environment.

The second respondents, the Nature Conservancy Council, did not appear B
and were not represented.

The following judgment was delivered.

SCHIEMANN J: This is a challenge to an order made under the Wildlife and Countryside Act 1981. By virtue of the provisions of Schedule 11 to that Act, if objection is duly made with respect to an order the Secretary of State for the
Environment must hold an inquiry, consider the report of his inspector and, if C
he decides to take no further action, then the order continues in effect. If any person is aggrieved by such an order and desires to question its validity on the ground that it is not within the powers of section 29 of the Act, then he may make an application to the court. If the court is satisfied that the order is not within the powers conferred by that section, then the court may quash the order. That is what I am asked to do.

D

Factual background

In 1985 Mr Sweet was notified by the Nature Conservancy Council (NCC) that they were of the opinion that Westhay Moor in Somerset was of special interest by reason of the flora, fauna or geological or physiographical features specified in the citation attached to the notification. The NCC notified him that they considered that the operations specified in the list attached to the
notification were likely to damage the flora or fauna or the features described. E
If one looks at the attached citation sheet one sees under the rubric "Description and Reason for Notification" the following: "Westhay Moor forms part of the nationally important grazing marsh and ditch systems of the Somerset Levels and Moors."

Among the voluminous reasons there given for that opinion, one finds the following: "Westhay Moor supports a nationally outstanding community of
terrestrial and aquatic vertebrates." The notification sets out a list of F
operations likely to damage the features of special interest.

On June 17 1987 the Secretary of State made the Westhay Moor (Somerset) Nature Conservation Order 1987. The relevant part of the order reads as follows:

> Whereas the land which is hereinafter described is in the opinion of the Secretary of State for the Environment of special interest and of national
> importance by reason of its grazing marsh, supporting a diverse and G
> abundant meadow and aquatic flora and fauna. And whereas it appears to the Secretary of State expedient to apply section 29(3) of the Wildlife and Countryside Act 1981 to the said land for the purpose of conserving those marshlands and its associated flora and fauna. And whereas the Secretary of State has consulted the Nature Conservancy Council.
>
> Now therefore the Secretary of State in exercise of the power conferred by
> section 29(1) and (3) of the said Act of 1981 hereby orders as follows: H
>
> (i) Section 29(3) of the said Act of 1981 shall apply to the land described in Schedule 1 hereto and shown hatched on the map contained herein.
>
> (ii) The operations specified in Schedule 2 hereto appear to the Secretary of State to be likely to destroy the marshlands habitat and associated flora and fauna.

Schedule 1 identifies three fields belonging to Mr Sweet and numbered 3300, 3192 and 2983, each being broadly rectangular and roughly the same size, 3300 being the northernmost rectangle, 2983 being the southernmost rectangle and 3192 being placed between them. They constitute one block of land on the map. The western and eastern sides of this block constitute ditches and there is also a ditch running along the northern edge of 2983 which separates it from 3192 and which joins the ditches running on the east and west sides of the order land.

Some 20 operations are specified in Schedule 2, including the following to which objection is taken on the basis that they are not properly called "operations":

> 1. Cultivation, including ploughing, rotavating, harrowing and reseeding.
> 2. Grazing.
> 3. Mowing or other methods of cutting vegetation.
> 4. Application of manure, fertilisers and lime.
> . . .
> 7. Burning.
> 8. The release into the site of any wild, feral or domestic mammal, reptile, amphibian, bird, fish or invertebrate, or any plant or seed.
> . . .
> 18. The storage of materials.
> . . .
> 20. The use of vehicles or craft likely to damage or disturb features of interest.

It is of interest that each of those activities, to use a neutral word, appears in the notification of the site of special scientific interest which was made in December 1985. The effect of such a notification under section 29 of the 1981 Act is to make it illegal to carry out those operations, but that illegality is only temporary. If the NCC are faced with an owner with whom they cannot reach agreement and yet they wish to have permanent protection for an area of special scientific interest, then they must proceed under section 29. The order under challenge was made under this section but, it is submitted, the Secretary of State went outwith the powers given by the section. It is therefore desirable that I should set it out:

> *Special protection for certain areas of special scientific interest.*
>
> **29.**—(1) Where it appears to the Secretary of State expedient to do so —
> (*a*) in the case of any land to which this paragraph applies, for the purpose of securing the survival in Great Britain of any kind of animal or plant or of complying with an international obligation; or
> (*b*) in the case of any land to which this paragraph applies, for the purpose of conserving any of its flora, fauna, or geological or physiographical features,
>
> he may, after consultation with the Nature Conservancy Council, by order apply subsection (3) to that land; and the provisions of Schedule 11 shall have effect as to the making, confirmation and coming into operation of orders under this section.
>
> An order made under this section may be amended or revoked by a subsequent order so made.
>
> (2) Paragraphs (*a*) and (*b*) of subsection (1) apply to any land which in the opinion of the Secretary of State is —
> (*a*) of special interest; and
> (*b*) in the case of paragraph (*b*) of that subsection, of national importance,
>
> by reason of any of its flora, fauna, or geological or physiographical features.
>
> (3) Subject to subsection (4), no person shall carry out on any land to which this subsection applies any operation which —
> (*a*) appears to the Secretary of State to be likely to destroy or damage the flora, fauna, or geological or physiographical features by reason of

which the land is land to which paragraph (*a*) or, as the case may be, paragraph (*b*) of subsection (1) applies; and A

(*b*) is specified in the order applying this subsection to the land.

(4) Subsection (3) shall not apply in relation to any operation carried out, or caused or permitted to be carried out, by the owner or occupier of the land if —

(*a*) one of them has, after the commencement date, given the Council notice of a proposal to carry out the operation, specifying its nature and the land on which it is proposed to carry it out; and B

(*b*) one of the conditions specified in subsection (5) is fulfilled.

(5) The said conditions are —

(*a*) that the operation is carried out with the Council's written consent;

(*b*) that the operation is carried out in accordance with the terms of an agreement under section 16 of the 1949 Act or section 15 of the 1968 Act[*] and;

(*c*) subject to subsections (6) and (7), that three months have expired since the giving of the notice under subsection (4). C

(6) If before the expiration of the period mentioned in paragraph (*c*) of subsection (5) the Council offer to enter into an agreement for the acquisition of the interest of the person who gave the notice under subsection (4) or an agreement under section 16 of the 1949 Act or section 15 of the 1968 Act providing for the making by them of payments to that person, that paragraph shall have effect as if for the said period there were substituted — D

(*a*) where the agreement is entered into before the expiration of twelve months from the giving of the notice, the period expiring on the day on which it is entered into;

(*b*) in any other case, twelve months from the giving of the notice or three months from rejection or withdrawal of the offer to enter into the agreement, whichever period last expires.

(7) If before the expiration of the period mentioned in paragraph (*c*) of subsection (5), or that paragraph as it has effect by virtue of subsection (6), an order is made for the compulsory acquisition by the Council of the interest of the person who gave the notice under subsection (4), that paragraph shall have effect as if for the said period there were substituted the period expiring — E

(*a*) in the case of an order which is confirmed, on the day on which the Council enter on the land;

(*b*) in any other case, on the day on which the order is withdrawn or the Secretary of State decides not to confirm it. F

(8) A person who, without reasonable excuse, contravenes subsection (3) shall be liable —

(*a*) on summary conviction, to a fine not exceeding the statutory maximum;

(*b*) on conviction on indictment, to a fine.

(9) It is a reasonable excuse in any event for a person to carry out an operation if — G

(*a*) the operation was authorised by a plannng permission granted on an application under Part III of the Town and Country Planning Act 1971 or Part III of the Town and Country Planning (Scotland) Act 1972; or

(*b*) the operation was an emergency operation particulars of which (including details of the emergency) were notified to the Council as soon as practicable after the commencement of the operation.

H

Pursuant to the procedures set out in Schedule 11, an inquiry was duly held before an inspector, who described the site as follows:

The Order land comprises three fields (OS 3300, 3192 and 2983) and the ditches/drains which lie between and which bound these fields together with

*The 1949 Act: the National Parks and Access to the Countryside Act 1949. The 1968 Act: the Countryside Act 1968.

the contiguous length of Paron's Drove in total 7.57 hectares, within and adjacent to the western edge of Westhay Moor SSSI.

The inspector made the following findings of fact:

> 1. The Order land includes three meadows and associated ditches totalling 7.57 hectares within the Westhay Moor SSSI.
>
> 2. The SSSI (463 hectares) is one of ten such areas notified within the Somerset levels and moors which generally is the most important grazing habitat left in Britain.
>
> 3. The special interest and national importance of Westhay Moor relates to both fauna and flora, ie to the presence of herb-rich grassland, ditch plant communities, terrestrial and aquatic invertebrates, breeding and wintering bird population and otters.
>
> 4. The Order land includes 4.45 hectares of herb-rich mire type grassland.
>
> 5. On the basis of the diversity of plant communities and rarity of species *this part of the Order site*

my emphasis

> is within the highest possible botanical classification of meadows.
>
> . . .
>
> 10. The ditches within the Order land are linked to the ditch systems of Westhay Moor.
>
> 11. The Westhay Moor SSSI is the most important in vertebrate area (*sic*) of the Somerset levels and moors in Britain.
>
> 12. Limited survey of the Order land ditches in 1987 found 61 species of aquatic invertebrates including three red data book and five nationally notable species.
>
> . . .
>
> 14. Because of the invertebrate community present the ditches of the Order site could be notified as an SSSI in their own right.
>
> 15. This is a mollusc based community unique in Britain.
>
> . . .
>
> 17. The Order land is unique in its combination of herb-rich mire grassland with ditch habitats.
>
> 18. Loss of either meadows or ditches would destroy the habitat.
>
> 19. Mr Sweet does not dispute the special interest and national importance of the fauna and flora of the Order land.
>
> . . .
>
> 27. Field 3192 was ploughed and partly cropped for potatoes about four to five years ago but the crop failed and the land was subsequently treated with Aldrin and has been fallow since.
>
> . . .
>
> 30. Mr Sweet's proposals to remove peat from field 3192 and spread it over the other two fields would lead to destruction of the present grassland sward.
>
> . . .
>
> 32. Turf production would be at the expense of the existing meadow habitat, eliminate the invertebrate interests of the site and disturb breeding and wintering birds. The red data book species of invertebrate would disappear and the nationally notable species would also be likely to be lost.
>
> 33. The previous use of Aldrin threatens the otter and invertebrate populations unless there is very careful management of field 3192 over a long period in the future.

The inspector's conclusions include the following:

> 47. In my opinion there can be no doubt that the Order land is of special interest and national importance both for its flora and fauna. The sole objector, Mr Sweet, does not in fact challenge that conclusion. The Somerset levels and moors generally are the most important grazing marsh habitat remaining in Britain. Westhay Moor is one of ten SSSI within the moors and levels. It has special interest and national importance by reason of herb-rich grassland, ditch fauna and flora, breeding and wintering bird populations and otters. The Order site is an integral part of Westhay Moor but is also itself

> significant for 4.45 hectares of herb-rich mire grassland, the invertebrate **A** community of the ditches and the relationship of ditch and grassland habitats.
>
> 48. Herb-rich mire grassland is a scarce resource nationally and has particular scientific interest as the interface between true mire and true grassland communities. The 4.45 hectares of this habitat within the Order land comprise a significant part of its total extent within Westhay Moor and within the Somerset levels and moors. The disturbance of field 3192 by **B** ploughing some years ago and the maintenance of a mown strip on field 2983 have not diminished the conservation significance *of the rest of the site*

my emphasis

> none the less the future management of these parts of the lands needs to be compatible with that required for the maintenance *of the nationally important parts of the Order area*

C

my emphasis.

> 49. The Order land ditches are part of the ditch system of Westhay Moor which is one of the most important areas in Britain for invertebrates. The significance of the aquatic invertebrate community of the Order land would in fact justify SSSI status for the ditches in their own right.
>
> 50. The ditches and the meadows each thus provide habitats of special interest and national importance but they are of course inter-related and **D** interdependent. Were either habitat not to be managed appropriately to achieve conservation objectives, demise of the one would lead to the disappearance of the other. In other words the Order land despite its different habitats and despite the different management given to fields 3192 and part of 2983 in the past constitutes a single environment and must be treated as such.
>
> . . .
>
> 54. NCC expressed particular concern at the inquiry when it was revealed **E** that Aldrin had been used on field 3192 some years ago. The evidence put to me does not suggest that there has been any serious detriment as yet to the flora and fauna of special interest and national importance on the site as a whole. Clearly, however, if the land has been treated with this pesticide the need for appropriate management in the future is emphasised. In other words, the Order land has to be managed in a way which conserves the special interest and national importance irrespective of the use of Aldrin and the **F** effect of use of that pesticide will be to vary the methods of management which are required.
>
> . . .
>
> 57. The evidence as to the scientific value of the Order land in my view amply demonstrates its special interest and national importance; this evidence and this conclusion was not challenged by the objector. I also consider that the special interest and national importance of the flora and fauna were threatened at the time that the Order was made and that continued **G** protection is necessary.
>
> 58. I recommend therefore that no action be taken to amend or revoke the Order.

The Secretary of State, having received this report dated December 7 1987, gave his decision with commendable celerity on January 21 1988 and stated:

> The Secretary of State notes the Inspector's findings of fact and agrees with **H** his conclusion. The Secretary of State remains satisfied that the Order land is of special interest and of national importance by reason of its marshland habitat which supports a diverse and abundant and aquatic flora and fauna and it appears to the Secretary of State expedient for the purpose of conserving that flora and fauna to continue to apply section 29(3) to that land.

Before turning to the contentions of the parties, it is right to note that Mr

A Sweet appeared in person before the inspector and made none of the points which counsel now makes on his behalf. The NCC were represented by counsel and called a Dr Johnson, who produced, it seems, a plan which identified the three fields and enclosed with a thick black line the major part of field 3300 and the major part of field 2983, writing within the enclosure the words "nationally important". A look at this plan helps one to understand the inspector's conclusions.

B Mr Howell submitted on behalf of the applicant that the Secretary of State had no power to make land the subject of the order which was not, in his opinion, of national importance by reason of its flora and fauna on the ground that it was desirable such land should be managed together with land which was, in his opinion, of national importance for such reasons:

(1) Only land which is of *national* importance by reason of *its* flora, fauna or geological or physiographical features could in this case be included within the C order land by virtue of section 29(2) of the Act.

(2) Land in OS 3192 and part of OS 2983 was not found to be of national importance by reason of *its* flora and fauna but was included only for the purpose of preserving and conserving the flora and fauna of *other* land within the order: see also section 29(1)(*b*).

(3) The Nature Conservancy Council have other powers to enter agreements D to secure that land is managed for the purpose of preserving flora, fauna or geological or physiographical features of special interest in the area where that appears to be expedient in the national interest and to acquire land compulsorily for that purpose: see sections 15, 16 and 17 of the National Parks and Access to the Countryside Act 1949 (as amended). Similarly, it has other powers to enter into agreements in respect of land which is or forms part of an area which in the NCC's view is of special interest by reason of its flora, fauna E or geological or physiographical features for the purpose of conserving them: see section 15 of the Countryside Act 1968.

(4) Section 29 should, he submitted, be given a narrower construction in case of ambiguity, given that nature conservancy orders curtail a landowner's rights and may lead to criminal proceedings and compulsory purchase.

He went on to submit that the order specifies matters in Schedule 1 that are not operations within the meaning of section 29(3) of the Act and elaborated F that argument as follows: (1) an order may only specify an "operation" carried out on the land, to which the order applies, that is likely to have certain effects: section 29(3); (2) an operation is an activity which results in some physical alteration to the land, which has some degree of physical permanence to the land itself. It does not embrace all activities on the land, such as the use made of it, which do not have such an effect: see section 29(9)(*a*) of the Act and *Parkes* v *Secretary of State for the Environment* [1979] 1 All ER 211 at G pp 213H–214G. The activities involved in agriculture, for example, are not, he submitted, all operations.

Then he pointed specifically to items (1) to (4), (7), (8), (18) and (20) in Schedule 2 to the order and submitted that these were not properly called operations.

In elaboration of his first submission, Mr Howell drew a distinction H between the ditches and the 4.45 hectares of herb-mire grassland, on the one hand, and on the other the three or so hectares which constituted the balance of the order land. He accepted that the inspector had found the former to be of national importance in its own right but, relying on those parts of the report which I have emphasised, submitted, and in my view rightly, that had the remaining three hectares been looked at on their own, then they would not have been found to be of national importance.

However, in my judgment the inspector's finding that loss of either meadows or ditches would destroy the habitat and his conclusion in para 50 that the whole of the order land constituted a single environment and must be treated as such was one to which he was entitled to come on the evidence. Indeed the contrary was not seriously argued in front of me or the inspector. In the light of the applicant's express acceptance in front of the inspector that the order land was of special interest and national importance, it is not surprising that the inspector so found without giving more reasons than he gave.

If the inspector was entitled, on the evidence before him, to come to the conclusion in para 50, then in my view those conclusions do not incorporate or betray an error in statutory interpretation. In my judgment, the Secretary of State was entitled to draw the boundaries of the order site where he did and treat the site as a whole and to find that it was of national importance by reason of its flora etc.

I can quite see that problems can occur where, for instance, rare animals graze on vegetation in one field, which vegetation is dependent for its continued life on an aquifer a mile away. I prefer to express no concluded view on such a case. The present case, however, presents no such problems and the challenge on the first ground fails.

The second ground of challenge was again one not advanced before the inspector and so one cannot be surprised that his report does not deal with it. In principle I approach the granting of relief on judicial review with hesitation if an applicant had the order under attack before him, appealed against it but did not then on the appeal take the points which he seeks to take for the first time before the court.

Mr Howell's attractive argument was based on: (1) the submission that the Act in section 32 uses the word "activities" when that is what is meant; (2) the submission that the carrying out of operations is a term used in the planning legislation which has been subject to much case law; (3) the common-sense assertion that walking over land of this nature or releasing a dog or goat over it can destroy rare fauna and one would not ordinarily describe such things as the carrying out of operations.

These points have force. However, "operations" is not a word of very precise meaning. It *is* used as a term of art in the planning Acts, where the carrying out of operations is contrasted with the making of a material change of use: see Town and Country Planning Act 1971, section 22, and the *Parkes* case (*supra*). I do not find planning law of any assistance in construing the statute with which I am concerned. I do not consider that the reasonable excuse referred to in section 29(9) of the 1981 Act is provided only if planning permission had been given for the carrying out of an operation and is not available if planning permission has been granted for the making of a material change of use.

Looking at the purpose of the Act as a whole, I consider the phrase in section 29(3) "An operation which appears to the Secretary of State to be likely to destroy the flora" etc wide enough to cover the matters specified in the Schedule and I am not prepared to find that the Secretary of State misconstrued the ambit of his powers. This application is therefore refused.

Application dismissed with costs.

R *v* Richmond upon Thames London Borough Council, *ex parte* McCarthy & Stone (Developments) Ltd

QUEEN'S BENCH DIVISION
Popplewell J
January 30 1989

Local planning authority — Consultations with prospective applicants for planning permission — Charge for consultations — Decision not to revoke
A charge — Whether charge *ultra vires* — Meaning of section 111 of the Local Government Act 1972

In July 1985 the respondent local planning authority resolved to impose
charges where certain members of the public sought consultations with their
officers. Representatives of the applicant company met with officers of the
respondent local planning authority for consultations prior to the submission
B of an application for planning permission. The respondents charged £25 for
each consultation session. The applicants requested the respondents to revoke
their decision to charge for these consultations; on October 27 1987 they
decided not to revoke their policy of charging for certain consultations. The
applicants challenged the decision of the respondents not to revoke the
charges; they claimed that there was no lawful authority to make the charges
because: (1) the powers in section 111(1) of the Local Government Act 1972
are exclusive and only empower anything which involves "expenditure,
C borrowing or lending of money", and in the absence of express mention of
raising money by charges, there is no power; the proviso in section 111(3) is
inclusive, covering any raising of money, and only authorises the raising of
money if there is other legislative power; and (2) the decision to impose the
charges was unreasonable so far as the *Wednesbury* principles were
concerned.

D *Held* The application was refused.

1. Section 111(1) of the Local Government Act 1972 provides that ". . . a
local authority shall have power to do any thing . . . which is calculated to
facilitate, or is conducive or incidental to, the discharge of any of their
functions". This subsection gives the necessary power to charge for pre-
planning application discussions or consultations. The limitation, in
subsection (3), that "a local authority shall not . . . raise money except in
accordance with" the enactments relating to rates or precepts refers only to
E raising money by means of rates, precepts or borrowing, or lending and does
not prevent the raising of money by the planning consultation charges. That
subsection is not exclusive, and the making of the charges does not have to be
subject to any other enactment: see p 27B-D.

2. There were no grounds for deciding that the respondents had made a
decision no reasonable authority could make; the minutes of the appropriate
committee meeting that authorised the charges were available, and these
F showed a situation that was far from *Wednesbury* unreasonableness: see
p 28C.

Cases referred to in the judgment

Associated Provincial Picture Houses Ltd v *Wednesbury Corporation* [1948] 1 KB 223; [1947] 2 All ER 680; (1947) 45 LGR 635, CA

Attorney-General v *Fulham Corporation* [1921] 1 Ch 440; (1921) 90 LJ Ch 281; 85 JP 213; 19 LGR 441

Attorney-General v *Wilts United Dairies Ltd* (1921) 37 TLR 884

Application for judicial review A

This was an application for judicial review of the decision of the respondents, Richmond upon Thames London Borough Council, that a decision to impose planning consultation charges should not be revoked.

Richard Rundell and Simon Bird (instructed by Metson Cross & Co) appeared for the applicants.

David Mole (instructed by the solicitor to Richmond upon Thames London Borough Council) appeared for the respondents. B

The following judgment was delivered.

POPPLEWELL J: The applicant in this case challenges the respondents' decision on October 27 1987 not to revoke their policy, that had been adopted on July 2 1985, namely, that a charge of £25 including VAT be made for enquiries relating to speculative redevelopment or development proposals C made by prospective purchasers of land or property.

On August 13 1986 the respondents charged the applicant £25 for a meeting, to be arranged with a planning officer, to discuss proposals to develop sheltered housing for the elderly at Mortlake Bus Garage, Avondale Road, Richmond. On January 12 1987 a further fee of £25 was charged for another meeting to discuss the revised proposals. D

The applicant contends that the respondents have no statutory authority to levy charges in respect of matters arising prior to the making of an application for planning permission. There is no dispute in this case that if a charge can properly be made there is nothing excessive about the sum of £25. The issue is whether the respondent council may lawfully charge for pre-application consultations with the council's planning officers.

As I have indicated, they were initially authorised by the council on July 2 E 1985. There was considerable correspondence between the parties as to the legality of that, and the respondents were invited to revoke that policy. They decided not to do so on October 27 1987.

Local authorities, it is agreed, are creatures of statute and can do only what a statute requires them to do or permits them to do; they may levy a charge only when expressly authorised by statute so to do. There is no dispute that it is for the authority to identify the statutory powers upon which they rely. F

A number of authorities were cited to the court, but it is necessary to make reference to only two. First, the case of *Attorney-General* v *Wilts United Dairies Ltd* (1921) 37 TLR 884.

The headnote reads:

> Where under the Defence of the Realm Acts and Regulations the Food Controller imposes as a condition of the grant of a licence to purchase milk in G certain areas a charge of 2d per gallon payable to him by the purchaser, the charge being part of a scheme for the regulation of prices, such charge was not within the powers of the Food Controller. Judgment of Bailhache J . . . reversed.

Scrutton LJ, at p 885, said:

> It is conceivable that Parliament, which may pass legislation requiring the H subject to pay money to the Crown, may also delegate its powers of imposing such payments to the Executive, but in my view the clearest words should be required before the courts hold that such an unusual delegation has taken place. As Chief Justice Wilde said in *Gosling* v *Veley* 12 QB at p 407: "The rule of law that no pecuniary burden can be imposed upon the subjects of this country, by whatever name it may be called, whether tax, due, rate or toll, except upon clear and distinct legal authority, established by those who seek

> to impose the burthen, has been so often the subject of legal decision that it may be deemed a legal axiom, and requires no authority to be cited in support of it."

Atkin LJ (as he then was), at p 886, said:

> In these circumstances, if an officer of the executive seeks to justify a charge upon the subject made for the use of the Crown (which includes all the purposes of the public revenue), he must show, in clear terms, that Parliament has authorised the particular charge. The intention of the Legislature is to be inferred from the language used, and the grant of powers may, though not expressed, have to be implied as necessarily arising from the words of a statute; but in view of the historic struggle of the Legislature to secure for itself the sole power to levy money upon the subject, its complete success in that struggle, the elaborate means adopted by the Representative House to control the amount, the conditions and the purposes of the levy, the circumstances would be remarkable indeed which would induce the court to believe that the Legislature has sacrificed all the well-known checks and precautions, and, not in express words, but merely by implication, had entrusted a Minister of the Crown with undefined and unlimited powers of imposing charges upon the subject for purposes connected with his department.

To like effect is what Sargant J said in the case of *Attorney-General* v *Fulham Corporation* [1921] 1 Ch 440. In that case the metropolitan borough of Fulham was minded to institute a change in the arrangements for those minded to bring their clothing for washing; it involved expenditure by those who brought their clothes. At p 449 Sargant J said:

> This is an action by the Attorney-General, at the relation of the ratepayers of Fulham, against the municipal borough of Fulham, to restrain it, to use a short phrase, from carrying on a laundry, or something in the nature of a laundry.
>
> In considering what the corporation may do, and what it may not do, I have to take as my guiding authority the words of Lord Selborne LC in *Attorney-General* v *Great Eastern Ry Co* (1880) 5 App Cas 473 at p 478. He says: "I assume that your Lordships will not now recede from anything that was determined in *Ashbury Ry Co* v *Riche* (1875) LR 7 HL 653. It appears to me to be important that the doctrine of *ultra vires*, as it was explained in that case, should be maintained. But I agree with James LJ that this doctrine ought to be reasonably, and not unreasonably, understood and applied, and that whatever may fairly be regarded as incidental to, or conseqential upon, those things which the Legislature has authorised, ought not (unless expressly prohibited) to be held, by judicial construction, to be *ultra vires*." That recognises that in every case it is for a corporation of this kind to show that it has affirmatively an authority to do particular acts; but in applying that principle, the rule is not to be applied too narrowly, and the corporation is entitled to do not only that which is expressly authorised but that which is reasonably incidental to or consequential upon that which is in terms authorised. And it is, of course, for the defendant corporation to point out the authority under which it has acted in what it has done.

There is no dispute as to the position of the local authority. The respondents are a local planning authority under section 1 of the Town and Country Planning Act 1971, as amended by the Local Government Act 1985, and as such are obliged by law to determine planning applications properly made to them. That appears from section 29 of the Town and Country Planning Act 1971.

By section 87 of the Local Government, Planning and Land Act 1980, the Secretary of State for the Environment has provided for the payment of fees in respect of, *inter alia*, planning applications. Section 87 reads:

> (1) The Secretary of State may by regulations make such provision as he A
> thinks fit for the payment of a fee of the prescribed amount to a local planning authority in England or Wales or a planning authority in Scotland in respect of an application made to them under the planning enactments for any permission, consent, approval, determination or certificate.

The regulations that were made were the Town and Country Planning (Fees for Applications and Deemed Applications) Regulations 1983 (SI 1674). Part
II of Schedule 1 laid down various fees, running up into the thousands of B
pounds, for various applications.

The Government issues circulars from time to time in relation to planning matters, and Circular 19/85 is in relation to an amendment to those regulations. It sets out fees for various particular schemes and also sets out various exemptions to those fees, of which disabled persons, permitted
development and revised applications are examples. C

It is quite clear that it is in everybody's interest that there should be discussions before parties submit their full application. There should be discussions between those who propose to carry out these development works and the local authority, so that everybody knows whether the scheme is likely to be successful or not. That has been set out in a number of circulars: Circular 22/80, Circular 28/83 and Circular 2/87. Circular 28/83, starting at para 8,
says: D

> Local authorities' attention is drawn to the need for:
>
> . . .
>
> (iv) early informal discussions with applicants and their agents to be encouraged. Where possible, these should include discussion about features of schemes which may give rise to the imposition of conditions in the event of permission being granted. Applicants can
> then consider the scope for adjusting the scheme prior to formal E
> submission of a planning application so as to render the conditions unnecessary if the local planning authority are minded to grant planning permission;
>
> . . .
>
> 9. The attention of applicants is drawn to the need for early discussions to be held with local planning authorities so that:
>
> (i) guidance can be given on how local authorities' planning policies F
> should influence the applicant's proposal before a formal planning application is submitted;
>
> (ii) the information required by the local authority for proper consideration of the application can be agreed, and provided with the formal application;
>
> (iii) applications are submitted in the correct form; and
>
> (iv) the correct fee is submitted with the application.
>
> . . . G
>
> 16. There will be some financial and manpower costs associated with the implementation of this code. These costs will vary between local authorities according to the extent to which information is already collected by them for their own use. However, there may be some offsetting benefits to be obtained by local authorities from the use of the information required by the code in the management of their development control duties.

H

The applicant says that that is an indication that the minister is aware that local authorities will incur costs in this exercise but has made no express provision to deal with it.

It is agreed on both sides that the solution to this problem lies in the construction of section 111(1) of the Local Government Act 1972. It is headed: "Miscellaneous Powers of Local Authorities, *Subsidiary powers*" and then "Subsidiary powers of local authorities":

> (1) Without prejudice to any powers exercisable apart from this section but subject to the provisions of this Act and any other enactment passed before or after this Act, a local authority shall have power to do any thing (whether or not involving the expenditure, borrowing or lending of money or the acquisition or disposal of any property or rights) which is calculated to facilitate, or is conducive or incidental to, the discharge of any of their functions.
>
> . . .
>
> (3) A local authority shall not by virtue of this section raise money, whether by means of rates, precepts or borrowing, or lend money except in accordance with the enactments relating to those matters respectively.

It will be seen that nowhere in that section is there any mention of the local authority raising money by making a charge to an individual or a company. What is said on behalf of the applicant, under subsection (1), is that the phrase which is contained in the part which is bracketed is exclusive; that is to say, the local authority have power only to do anything which involves expenditure, borrowing or lending of money, or the acquisition or disposal of any property or rights under that section. The absence of any mention of raising money by charging an individual supports the argument that there is a very great limitation on the local authority's power under that section. That, on initial reading, has considerable force.

However, it is pointed out that under subsection (3) a local authority shall not raise money, "whether by means of rates, . . . except in accordance with the enactments relating to those matters respectively". What the local authority say is that it is clearly intended that the powers in subsection (1) contemplate the raising of money by rates, otherwise it would be quite unnecessary for that phrase to appear in subsection (3). What subsection (1) means, in relation to the matters contained in the bracketed part, is that those matters are inclusive of what the local authority have power to do, but are not exclusive; they are but examples. That is supported, it is said, by the fact that unless that were so there would be no need for subsection (3) to mention raising money by means of rates, precepts or borrowing.

I think, after argument, that counsel for the applicant accepted that construction, though it was not his initial argument. However, the argument then turns to what is meant by subsection (3). The local authority say that what subsection (3) does is to limit them so far as the raising of money by way of rates, precepts, or borrowing or lending is concerned. They are governed by those enactments which relate to raising money by way of rates, precepts or borrowing. The phrase "relating to those matters respectively" means that you have to look at other statutory authority for the means of rates, precepts or borrowing.

They say that the combined effect of section 111(1) and section 111(3) is this: that a local authority under section 111(1) shall have power to raise money, provided it is calculated to facilitate, or is conducive or incidental to, the discharge of any of their functions, but that so far as raising money by rates, precepts or borrowing is concerned, that may not be done unless other enactments allow it to take place.

It is not, I think, seriously contended that the raising of money by charging an individual (or company in this case) is other than calculated to facilitate, or is conducive or incidental to, the discharge of any of the local authority's functions.

What the applicant says in relation to section 111(3) is that the phrase "raise money" is a broad term. It covers what has happened in this case, and by subsection (3) a local authority are not entitled to raise money unless there is some other enactment which entitles them so to do; there being, it is conceded,

no other enactment which entitles them to do, the local authority have no A
power to do it.

The phrase "whether by means of rates, precepts or borrowing" is, says the applicant, not exclusive; it is inclusive. So that "raising money" is broad, and the phrase "whether by means of rates, precepts or borrowing" is simply giving examples. Therefore, says the applicant, there being no power to raise money in this way, the local authority have acted illegally.

In my judgment, section 111(1) gives the local authority power to raise B
money by charging an individual, or company, for discussion about preplanning applications. It is calculated to facilitate, or is conducive or incidental to, the discharge of any of their functions.

By subsection (3), the raising of money by charging in this way is not excluded and is not subject to some other Act. The limitation on the raising of
money in subsection (3) refers only to raising money by means of rates, C
precepts or borrowing, or lending money. It is not contended that the raising of money in this case is by means of rates, precepts or borrowing. Accordingly, in my judgment the local authority do have power to do that which they have done, and the challenge to the legality of it therefore fails.

There is a second limb to the argument on behalf of the applicant which was an alternative, and made plainly so, namely, that if the power does exist, the
decision to impose it was *Wednesbury* unreasonable. Although strictly this is D
an application in relation to a refusal to revoke, I think it is necessary to look at the way in which the decision was made. It arises from a meeting of the planning committee on July 2 1985. The minutes of that meeting are before the court and read as follows:

> As a result of the investigations that have taken place to generate further
> income to the Council, a suggestion has been made that a charge be levied in E
> respect of certain categories of planning information that is given by the staff. This matter has now been investigated and is considered to be practical, the enquiries falling into two categories:
>
> 1. General enquiries which are made on behalf of prospective purchasers of land or property. These are often made through the normal Land Search procedure, whereby a charge of £1.80 per enquiry is levied. There would
> appear to be a case for levying a similar charge when such enquiries are made F
> by letter to the Planning Division.

And a second category, germane to the present application:

> 2. Enquiries relating to speculative redevelopment or development proposals, again made usually by a prospective purchaser of land or property. This sort of enquiry often arises where land or property is placed on
> the market and several developers or prospective purchasers make enquiries G
> of the Planning Office as to future proposals, in order that a bid can be made for purchase. This type of case involves the staff in a considerable amount of time and research and a fee would seem to be appropriate. Enquiries are often received from owners of land who wish to establish a value for future sale of their property and again, as this is a speculative matter, a fee would seem to be reasonable. Where informal proposals are submitted prior to the making of a planning application the information given often results in a more realistic
> planning application being submitted and permission being subsequently H
> granted, thereby avoiding a refusal of permission. In these cases it is considered necessary to have a system which is simple and easy to operate and therefore a flat fee of £20 is considered reasonable. As regards casual enquiries made at the enquiry desk, information would still be given provided it did not involve considerable search and staff time. In those cases which could not be dealt with at the time requested then the enquirer would be asked to put their enquiry in writing and enclose the appropriate fee.

> It is not considered appropriate that a fee should be levied on occupiers who merely wish to obtain information for improving their premises for their own occupation and this would include both businesses and residents. Similarly, enquiries of a general nature made by Associations or Societies should not normally attract a fee.
>
> It is clearly very difficult to assess the likely income of such a scheme but a simple survey of such enquiries suggests that income of around £7,000 per annum could be realised which has been included in the Council's two-year Financial Strategy for 1985/86.

The attack on that report is based on a number of matters. The only matter drawn to the council's attention was the raising of the charge, and it failed to draw to the council's attention the level of application fees (the exceptions which are set out in the circular); the circulars themselves; the desirability of such meetings, and the effects such charges might have on potential developers. I conclude that it is very unlikely that a planning committee of the local borough were not fully aware of the circulars, of the desirability of such meetings (to which they do in fact refer in this discussion); or the exemptions.

It seems to me that this case is miles away from being *Wednesbury* unreasonable. The planning committee seem to me to have come to a thoroughly sensible decision. They have had before them sufficient material upon which to come to that conclusion. The suggestion that the decision was unreasonable is, in my view, itself unreasonable. Accordingly, this application is refused.

Application dismissed with costs. Leave to appeal granted.

White *v* Secretary of State for the Environment and another

COURT OF APPEAL
Mustill and Stuart-Smith LJJ and Sir John Megaw
February 1 1989

Use commenced before July 1 1948 — Existing use — Use abandoned — Established use certificate — Whether any distinction between existing use and established use — Whether concept of abandonment applies to existing use development A

In April 1982 the appellant applied for an established use certificate for
"winter storage for showmen and caravan site for showmen". In October
1982 the local planning authority, Congleton Borough Council, issued an
enforcement notice alleging a breach of planning control of the land in
question to use for the "residential occupation of caravans". The appellant's
appeals, against the refusal of the local planning authority to give the
certificate and against the enforcement notice, were dismissed by the B
Secretary of State for the Environment in April 1984. Appeals against those
decisions were dismissed by Mann J (March 12 1987).

The site was occupied from the early 1940s to 1969 by a travelling circus; in
1964 planning permission was granted for the temporary use and in 1971 a
certificate of established use was issued for the storage of caravans and circus
equipment during the winter months. The site then remained virtually
unoccupied between 1974 and 1981. However, the Secretary of State did not C
consider that the evidence established that the use of the site during the
currency of the 1964 temporary planning permission included any residential
caravan element independent of that permitted by that permission. It could
not be shown that there was a caravan site use in contravention of planning
control: an established use certificate could not therefore be issued at this
stage. The appellant contended that although the circus use ceased for a
number of years, it was not possible in law to abandon a use that had
commenced before July 1 1948, the appointed day: such a use was an existing D
use and not an established use.

Held The appeal was dismissed.

1. In *Pioneer Aggregates (UK) Ltd* v *Secretary of State for the Environment*
[1985] 1 AC 132, the House of Lords decided that a planning permission
enured for the benefit of the land and could not be abandoned. But that case
must be distinguished, as it did not concern either existing or established uses. E
The earlier case of *Hartley* v *Minister of Housing and Local Government*
[1970] 1 QB 413 decided that a use of land, whether an existing use or an
established use, could be abandoned: see pp 33H and 35H-36C.

2. It could not be argued that as the planning legislation was a
comprehensive code, and made no express provision for abandonment, the
concept had no application; the introduction of the legislation in 1948 left
existing uses outside the control of the Acts: see p 35H. F

3. Section 33 of the Town and Country Planning Act 1971, which provides
that planning permission enures for the benefit of the land and all persons
interested in it, has no application to existing uses: see p 34H.

4. Schedule 24, para 12, to the 1971 Act, which applies section 23(1) of the
Act (the necessity to obtain planning permission for development), only
excludes existing use development from the need for planning permission
where such development is "carried out"; it is implicit that the use is a
continuing use: see p 36C.

Cases referred to in the judgments

Barling (David W) Ltd v *Secretary of State for the Environment* [1980] JPL 594

Hartley v *Minister of Housing and Local Government* [1970] 1 QB 413; [1970] 2 WLR 1; [1969] 3 All ER 1658; (1969) 68 LGR 32; 21 P&CR 1; [1970] EGD 4; 213 EG 31, CA

Pioneer Aggregates (UK) Ltd v *Secretary of State for the Environment* [1985] 1 AC 132; [1984] 3 WLR 32; [1984] 2 All ER 358; (1984) 82 LGR 488; 48 P&CR 95; [1984] EGD 1094; 272 EG 425; [1984] JPL 651, HL

Appeal against a decision of Mann J

This was an appeal against a decision of Mann J (March 12 1987) in which he dismissed an application and appeal against decisions of the Secretary of State for the Environment to dismiss an appeal against a refusal of planning permission and to dismiss an appeal against an enforcement notice issued by Congleton Borough Council.

Barry Payton (instructed by Geoffrey T Smith & Co, of Wolverhampton) appeared for the appellant.

Robert Jay (instructed by the Treasury Solicitor) appeared for the first respondent, the Secretary of State for the Environment.

The second respondent, Congleton Borough Council, did not appear and was not represented.

The following judgments were delivered.

STUART-SMITH LJ: The principal question posed in this appeal is whether a use of land which existed before the coming into force of the Town and Country Planning Act 1947 on the appointed day, namely July 1 1948 (an existing use), can be abandoned, so that resumption of the use constitutes development for which planning permission is required.

The appeal is from the judgment of Mann J given on March 12 1987 in which he upheld the decision of the Secretary of State for the Environment contained in a decision letter dated April 18 1984 in which the Secretary of State held that, among other things, the pre-existing use had been abandoned and that its resumption required planning permission.

There were before Mann J two matters, the first an application under section 245(1) of the Town and Country Planning Act 1971 and an appeal under section 246(1) of that Act. The appellant is Mr Joe White, who has since 1981 had an interest in land known as the Old Brickworks, Newcastle Road, Brereton. The respondents are the Secretary of State for the Environment and the Congleton Borough Council. The borough council are the local planning authority for the area which includes the Old Brickworks. They did not appear before the judge or before this court.

On April 28 1982 the appellant applied for an established use certificate. The use for which the certificate was sought was "winter storage for showmen and caravan site for showmen". The borough council on June 30 1982 refused the application. The applicant appealed to the Secretary of State, who on April 18 1984 dismissed the appeal. That decision was the subject of the application under section 245 of the Act of 1971.

On October 27 1982 the borough council issued an enforcement notice. It alleged a breach of planning control and asserted that the breach was the use of the Old Brickworks for "the residential occupation of caravans". The steps required to be taken were the discontinuance of the use "for the residential occupation of caravans". The appellant appealed to the Secretary of State, who on April 18 1984 dismissed the appeal but amended the enforcement notice. That decision was the subject of the appeal under section 246(1) of the Act of 1971.

It is common ground that the fate of both the application and the appeal is governed by the same considerations.

I can take the facts from the judgment of the learned judge. The appellant's wish is that the site should, during the winter months, accommodate showground equipment and caravans; most importantly, caravans occupied by people who would maintain the machinery of the showground equipment and prevent damage by intruders who could cause damage either deliberately or accidentally; it is those caravans which occasion the present dispute.

There was an inquiry held on June 21 1983 by Mr N Barclay and on July 4 1983 Mr Barclay made his report to the Secretary of State. He made findings of fact which were adopted by the Secretary of State and recorded in the decision letter. Para 6 of that letter reads:

> The Inspector found as facts, which are accepted, that, originally a brickworks, the appeal site was occupied from the early 1940s to 1969 by a travelling circus and, on its departure to new quarters nearby, was left vacant until 1970 when motor vehicles were stored but later removed as a result of enforcement action in December 1970 alleging the dismantling and storage of vehicles and parts of vehicles. In 1971 an established use certificate was granted for the storage of caravans and circus equipment during the winter months, stating that the use existed on the site on 4 June 1971. In 1972 a residential caravan was stationed on the land and occupied as such for about 18 months: when this was removed in 1974 the premises remained vacant, except for some grazing and occasional unauthorised overnight parking of touring caravans, for approximately 7 years. In September 1981 the site was occupied by your client with his group of residential caravans, cars, lorries and trailers: they departed in February 1982 and returned with an augmented group in about October 1982, which left again in the spring of 1983 leaving one residential caravan and a box bodied van on the land. From the mid-1950s to the late 1970s the site was the subject of various planning applications for residential development, the stationing of caravans and commercial uses: all were refused, except that in 1964 permission was granted to the circus owners for the stationing of two residential caravans on the land for five years to December 1968. The permission limited the occupation of the caravans to the applicant, who was one of the circus proprietors, and persons employed by him: this person was a caretaker.

The critical paragraph in the decision letter is para 8, which reads:

> Taking first the use of the site during the 1960s, whilst it is accepted that the site was used for the stationing of residential caravans during that period until the circus moved elsewhere in 1969, it is not considered that the evidence establishes that the circus's use of the site during the currency of the temporary permission granted in 1964 included any residential caravan element independent of that permitted by that permission. It is therefore considered that at least insofar as your client's claim to an established use certificate is in respect of the use of the site as a winter caravan site for showmen, it must fail on the basis that it has not been shown that during the period 1964–1969 there was a caravan site use of the site in contravention of planning control (see *Bolivian and General Tin Trust* v *Secretary of State for the Environment* [1972] 1 WLR 1481). It is also considered that even if it were to be accepted that immediately prior to the departure of the circus in 1969 the site did have an established seasonal use extending not only to the storage of circus equipment but also including a residential element, having regard to the evidence as to the use of the site from 1969 onwards it can only be concluded that when in 1981 your client moved in, there was then no dormant winter quarters use available for him to resume. Whilst the evidence as to the history of the site from 1969 to 1981 is in some respects rather sketchy, it is considered that as against the pre-1969 use by the circus there would have been a material change of use with the introduction of the use for vehicle breaking and storage enforced against in 1970, and that the introduction of that use would rule out any possible claim to a right to resume the pre-1969

> "circus" use on the basis that that use had not been abandoned (see *Young* v *Secretary of State for the Environment and London Borough of Bexley*, CA, [1983] JPL 465). Additionally, there would appear to have been a further material change of use of the site with the introduction in circa 1972 of a residential caravan site use by Mr Barker's father-in-law. Further, even if notwithstanding these changes of use it remained relevant to consider whether the "circus" use, a use discontinued on the ground in 1969 when the circus moved out, had nevertheless not been abandoned by 1981, having regard to the length of the period during which that use was not physically being carried on, to the intervening uses, and to the period from 1974-1981 when the site was substantially vacant, it is considered reasonable to conclude that by 1981 that use had been abandoned.

The learned judge analysed the Secretary of State's reasons into four as to why a preappointed-day use had no materiality in 1981. (1) Prior to 1969, when the circus departed elsewhere, there was no residential use attributable to the use for storage as opposed to a use attributable to the temporary planning permission granted in1964. (2) Even if there had been such a use, then (a) vehicle processing was a material change of use whose introduction would rule out any reversion to the circus use; (b) even if that was not the case, there was in 1972 the material change of use when the site was devoted to a residential caravan use which was a material change of use whose introduction would rule out any reversion to the circus use; (c) in any event, it was reasonable to conclude that the circus use had been abandoned by 1981.

The ground (2)(a) was not pursued by or on behalf of the Secretary of State before the learned judge and has not been pursued before us. The learned judge upheld the decision of the Secretary of State on each of the other three grounds. But the argument before us is concentrated primarily on the ground in para (2)(c).

Accordingly, I propose to consider this ground first. Mr Payton's submission is that there can be no such thing as abandonment of an existing use. It is by no means clear whether this point was argued before the learned judge or whether the argument was in fact confined to the submission that on the facts the use had not been abandoned. The judge clearly thought that the latter was the issue and the notice of motion supports this view. Neither counsel who appeared before us was in the court below. Nevertheless, the point is raised in the notice of appeal, and Mr Jay on behalf of the Secretary of State took no point on it.

Mr Payton submitted that, since the decision of the House of Lords in *Pioneer Aggregates (UK) Ltd* v *Secretary of State for the Environment* [1985] 1 AC 132, there could be no such thing as abandonment of an existing use. To examine this submission it is necessary first to consider the case of *Hartley* v *Minister of Housing and Local Government* [1970] 1 QB 413, because it is the case upon which the judge founded his decision, it is binding upon us and Mr Jay submits that it decides the point in this case. The facts were that in 1961 F bought a site, part of which had been used as a petrol filling station and another part for the display and sale of cars. A few months later F became ill and died. His widow and young son carried on the petrol filling station business but gave up the sale of cars. In 1965 the appellant bought the site and used it both as a filling station and for the sale of cars. In 1967 the planning authority served an enforcement notice on him, alleging that the use of the site for display and sale of cars constituted development without grant of permission and requiring him to discontinue that use. The appellant appealed to the minister, who upheld the enforcement notice; he found that by 1965, if not in 1961, the use for car sales had been abandoned, and the use of the site was that of a petrol filling station only, and consequently the use for display

and sale of cars in 1965 was a material change of use. The appellant's appeals A
to the Divisional Court and Court of Appeal were dismissed. It is not entirely
clear from the facts whether the use for display and sale of cars was an existing
use, ie one that existed before July 1 1948, or was an established use, ie one that
started after that date but was immune from enforcement because no notice
had been served within four years. The facts as they appear in the report of the
case before the Divisional Court suggest that it was the former (see [1969] 2 QB
46 at p 52G). But it is clear that the Court of Appeal drew no distinction B
between the two for the purpose of the case and, as will appear, the House of
Lords in the *Pioneer Aggregates* case treated it as an existing use case.

Mr Glidewell QC (as he then was) argued that where there was a use of land
which was either an existing use on the appointed day or an established use
because it had continued for more than four years before service of the
enforcement notice, its use could not by mere cessation be abandoned such as C
to make its resumption development. Similarly, where there had been dual
use, followed by single use, and then a resumption of that use, there was no
abandonment of the second use by ceasing it for a time. This argument was
rejected; at p 420E Lord Denning MR said:

> I think that when a man ceases to use a site for a particular purpose and lets it
> remain unused for a considerable time, then the proper inference may be that
> he has abandoned the former use. Once abandoned, he cannot start to use the D
> site again, unless he gets planning permission: and this is so, even though the
> new use is the same as the previous one.
>
> The material time is when he starts on the new use. You have to ask at that
> time whether there was then a material change of use — from a non-use into a
> positive use. Take this very case. In 1965 Mr Hartley started to use the site for
> selling cars. That was a new use. It was a change from a non-use into a use for
> selling cars. It needs planning permission — unless Mr Hartley can say that E
> the previous use was never abandoned.
>
> The question in all such cases is simply this. Has the cessation of use (followed
> by non-use) been merely temporary, or did it amount to an abandonment? If
> it was merely temporary, the previous use can be resumed without planning
> permission being obtained. If it amounted to abandonment, it cannot be
> resumed unless planning permission is obtained. I said as much in *Webber* v
> *Minister of Housing and Local Government* [1968] 1 WLR 29, 33 and in
> *Miller (TA) Ltd* v *Minister of Housing and Local Government* [1968] 1 WLR F
> 992, 996. Abandonment depends on the circumstances. If the land has
> remained unused for a considerable time, in such circumstances that a
> reasonable man might conclude that the previous use had been abandoned,
> then the tribunal may hold it to have been abandoned.

Widgery LJ at p 421 said:

> It has been suggested in the courts before, and it seems to me that it is now G
> time to reach a view upon it, that it is perfectly feasible in this context to
> describe a use as having been abandoned when one means that it has not
> merely been suspended for a short and determined period, but has ceased
> with no intention to resume it at any particular time. It is pefectly true, as Mr
> Glidewell says, that the word "abandonment" does not appear in the
> legislation. We are not concerned with the legislation at this stage, but merely
> with the facts of the matter. I cannot think of a better word to describe a H
> situation in which the landowner has stopped the activities constituting the
> use not merely for a temporary period, but with no view to their being
> resumed. If that has happened, then as a matter of fact the use has ceased.

In *Pioneer Aggregates (UK) Ltd* v *Secretary of State for the Environment*
[1985] 1 AC 132 the House of Lords held that a grant of planning permission
enured for the benefit of the land and all persons for the time being interested
in it and consequently a valid planning permission capable of implementation

A could not be abandoned by the conduct of the owner or occupier of the land for the time being. The case concerned grants of planning permission and not existing or established uses of land. Lord Scarman, with whose opinion the other learned lords agreed, gave three main reasons for this conclusion. The first is what I might call the policy reason. At p 139D he said:

> If the board is right, a valid planning permission can be abandoned by the
> B conduct of a landowner or occupier of land; and the effect of the party's conduct will be to bind all persons interested in the land now or hereafter whether or not they have notice of the abandonment. The planning permission would be entered in a public register; but not so its abandonment. Nor would it be possible by inspection of the land to discover whether the permission has been abandoned, for the absence of implementation of a planning permission is no evidence that a valid permission does not exist. It is
> C perhaps not surprising that no trace of any such rule can be found in the planning legislation. If there be such a rule, it has been imported into the planning law by judicial decision.

As was pointed out in argument by Sir John Megaw, and accepted by Mr Payton, the present case is the antithesis of this. The only way that a stranger would know if there is an existing use is if it is continuing. No doubt temporary
D cessations will not destroy the existing use; but, if it has not been exercised for many years, it would be a serious trap for those purchasing neighbouring land if they were to find that a long extinct use was resurrected simply on the ground that it had existed before July 1 1948.

The second ground was that the Town and Country Planning Acts were a comprehensive code and there was no provision for abandonment. At p 141C Lord Scarman said:

E

> Parliament has provided a comprehensive code of planning control. It is currently to be found in the Town and Country Planning Act 1971, as subsequently amended. Part II of the Act of 1971 imposes upon local planning authorities the duty of preparing and submitting to the Minister development plans formulating their policy and their general proposals for the development and use of land in their area. Widespread publicity has to be
> F given to the preparation or alteration of such plans. There is provision for local public inquiries in certain specified circumstances. Part III imposes general planning control. Section 23(1) declares the rule: subject to the provisions of the section, planning permission is required for the development of land. There are certain exceptions, of which the most notable are rights in connection with the use of land existing prior to certain specified dates related to the introduction of planning control (commonly called
> G "existing use rights"); sections 23 and 94 of the Act. Section 29 deals with the grant of planning permission: note that the local planning authority must have regard to the provisions of the development plan. In determining an application for permission the authority must take into account "any representations" made to them within the time specified in the section. And there are extensive provisions for giving publicity to applications: sections 26 to 28.

H

In my judgment, this reason has no bearing on the present problem, since it is common ground that existing use cases are outside the control of the Acts, as Lord Scarman states in the passage quoted.

Finally, reliance was placed on section 33(1) of the Town and Country Planning Act 1971, which provides that planning permission to develop land shall enure for the benefit of the land and all persons for the time being interested in it. This ground also has no application to the present case.

Mr Payton places reliance on the following passage in Lord Scarman's speech at p 142G:

> Viewed as a question of principle, therefore, the introduction into the planning law of a doctrine of abandonment by election of the landowner (or occupier) cannot, in my judgment, be justified. It would lead to uncertainty and confusion in the law, and there is no need for it. There is nothing in the legislation to encourage the view that the courts should import into the planning law such a rule — recognised though it is in many branches of the private law (eg the law of easements, the commercial law, and the law of trade marks) as Megarry J in his learned, though obiter, discussion of the principle has shown in *Slough Estates Ltd* v *Slough Borough Council (No 2)*, 19 P&CR 326.

But, in my judgment, that passage must be read in the context of the case, namely that it was concerned with planning permission. That this is so is made clear from what follows immediately. At p 143A Lord Scarman said:

> There is, however, quite apart from the *Slough* case a number of reported judicial decisions which, upon first sight and before analysis, might seem to suggest that there is room in the planning law for a principle, or an exception, allowing the extinguishment of a planning permission by abandonment. Three classes of case can be identified. The first class is concerned not with planning permission but with existing use. In *Hartley* v *Minister of Housing and Local Government* [1970] 1 QB 413 the Court of Appeal (Lord Denning MR, Widgery and Cross LJJ) held that the Minister as the tribunal of fact was entitled to find on the evidence that the resumption of a car sales use on a site where previously there had been two uses, namely car sales and a petrol-filling station, was after a cessation of the car sales use for some four years a material change of use and so properly the subject of an enforcement notice. The Minister, the court held, was entitled to find as a fact that the previous use had ceased, having been abandoned by the owner or occupier of the land. This was not a case of abandoning a planning permission. There was in fact no existing use of the land for car sales because the use had ceased years ago. An existing use, which has been deliberately ended before a resumption arises, is not existing at the date of resumption: accordingly, the resumption was a material change of use, and so required planning permission. The issue was one of fact, as Widgery LJ emphasised in his judgment. And it had nothing whatever to do with the extinguishment of a planning permission. Widgery LJ in the course of his judgment made a significant comment, at p 422:
>
> > "When the car sales use ceased in 1961 there could be no question of a material change of use on which an enforcement notice could be founded in reliance on that fact alone."
>
> The use no longer existing, the change back four years later was the material change of use on which the notice could be founded.

I cannot accept Mr Payton's suggestion that Lord Scarman confused existing uses and established uses and that he meant to refer to the latter. Quite apart from the extreme improbability of such a thing, it is clear that Lord Scarman had well in mind what existing uses were, since he defined the term at p 141E-F. In this passage, in my judgment, Lord Scarman plainly treated *Hartley's* case as an existing use case and approved the decision.

Mr Payton accepts that *Hartley's* case was rightly decided on its facts, but he submits that it can be distinguished. He further submits that, in the light of Lord Scarman's opinion, the exceptions to the rule that the doctrine of abandonment has no place in planning law must be confined to the three expressly dealt with in Lord Scarman's speech, of which *Hartley's* case is the only relevant one.

The grounds upon which he seeks to distinguish *Hartley's* case are, first, that it was a case of established use and not existing use. As I have already said, the facts are not entirely clear; what is clear is that the Court of Appeal drew no distinction between the two for this purpose and Lord Scarman treated the case as an existing use case. There is nothing in this point.

Second, he submits that there may be an abandonment of one of two existing uses, so that subsequent resumption of one may be a material change of use, particularly if the use that continued throughout is inconsistent with the discontinued use. He submits that there is no such thing as a nil use. I cannot see any logical basis for this submission. It certainly forms no part of the reasoning of the Court of Appeal in *Hartley's* case and is quite inconsistent with the last-quoted passage from Lord Scarman's speech in the *Pioneer Aggregates* case. It is also worthy of note that in argument in *Hartley's* case Mr Glidewell QC's alternative argument was that, if there could be abandonment of an existing use, it was limited to cases where there was a single use, cessation amounting to a nil use.

Mr Payton's final argument was based on Schedule 24, para 12, to the Town and Country Planning Act 1971, which provides:

> Subsection (1) of section 23 of this Act applies (subject to the provisions of that section) to the carrying out of development whether before or after the commencement of this Act, except that it does not apply to development carried out on or before the appointed day.

On a literal construction of this paragraph, he submitted that, once an existing use was established, it was for all time outside the ambit of the planning Acts, and therefore a resumpton of such a use after cessation does not fall within section 23(1) of the Act at all. But it is implicit in the words "carried out", at least so far as a use of land is concerned, that it is a continuing use. And this is inherent in the decision in *Hartley's* case.

For these reasons, in my judgment, the principal ground of appeal fails.

The second reason why the learned judge upheld the Secretary of State's decision was para 2(b) in his analysis, namely that in 1972 there had been a material change of use when the site was devoted to a residential caravan site use.

The evidence was that from some time in 1972 there had been a residential caravan on the site for 18 months, occupied by the then owner's father-in-law. Mr Payton submits that, first of all, this was not a separate reason found by the Secretary of State but merely part and parcel of, and bound up with, the finding of abandonment. I do not accept this; in my judgment, the judge's analysis of the reasoning in para 8 of the decision letter was correct. Second, he submits that the occupation of one caravan on the site, which was more than an acre, was not inconsistent with the existing use of winter quarters for showground equipment coupled with caravans occupied by those who looked after them. But I accept Mr Jay's submission that, if indeed there was a use of the land for caravans occupied by those who looked after the equipment, which was over and above and independent of the limited permission in that behalf, it was ancillary to the main use, namely the storing of the equipment. If the principal use is discontinued, the ancillary use cannot survive as the sole or principal use without planning permission. *David W Barling Ltd* v *Secretary of State for the Environment and Swale District Council* [1980] JPL 594.

In any event, this appears to me to be a question of fact and there was material upon which the Secretary of State could reach his conclusion. This court cannot substitute its own view, even if it differed, unless it can be shown on well-known principles that the Secretary of State has erred. Understandably, Mr Payton has not attempted this task, but he submits that the case should be remitted to the Secretary of State for reconsideration. Since I reject Mr Payton's first submission under this head, I can find no reason to send the matter back to the Secretary of State. A

There has been little or no argument addressed to us on the third ground, numbered 1 in the judge's analysis, upon which both the Secretary of State and the judge rejected the appellant's case, namely that prior to 1969, when the circus departed elsewhere, there was no residential use attributable to the use for storage as opposed to a use attributable to the temporary planning permission granted in 1964. But, even if this ground were wrong, and I am certainly not prepared to say that it was, I would dismiss the appeal on the B other two grounds.

SIR JOHN MEGAW: I agree that this appeal should be dismissed. I agree fully with the reasons given by Stuart-Smith LJ.

MUSTILL LJ: I also agree that this appeal should be dismissed.

The appeal was dismissed. No order made for costs save legal aid taxation. C
Leave to appeal to the House of Lords was refused.

R *v* Rochester upon Medway City Council, *ex parte* Hobday

QUEEN'S BENCH DIVISION
Rose J
February 2 1989

Enforcement notice — Stop notice — Appeal against refusal of planning permission for market use — Commencement of market use — Decision to take enforcement proceedings at expiration of 14 days — Whether decision to
A issue enforcement notice before expiration of 14 days valid — Whether ratification of decision to issue enforcement notice by later committee meeting valid — Effect of committee meeting authorising enforcement notices

The applicant acquired a site in Chatham for use as a market if planning permission were granted. An appeal against the deemed refusal of planning permission was dismissed on September 6 1988. In fact the applicant started to
B use the land for market purposes on October 31 1987. On November 17 1987 the appropriate committee of the respondent local planning authority resolved that on or after the 15th day of the use of the site as an open air market, the appropriate enforcement notice be issued and served together with stop notices prohibiting the appropriate breaches of planning control. On March 30 1988 the committee noted that authorisation had already been given for enforcement action at their previous meeting and authorised as a
C matter of urgency that a stop notice be served. As a consequence, an enforcement notice was issued in respect of the alleged breach of planning control and it and a stop notice prohibiting a further breach were sent to the applicant on April 8 1988, a day on which the site had only been used for less than 14 days for market purposes in 1988. The applicant had been claiming the benefit of permitted development for market use on 14 days per calendar year in accordance with class IV of the Schedule to the Town and Country Planning
D General Development Order 1977. Following an application by the applicant for judicial review of that first stop notice, the respondents withdrew it and purported to serve a second stop notice dated April 14 1988. That second notice was suspended, following objections by the applicant, and on April 25 1988 the respondents resolved to serve a third stop notice, the subject of the present case.

The applicant contended that the decision of the committee on November
E 17 1987 was a nullity being outside the respondents' statutory powers: (1) because there was no breach of planning control at the time apparent to the respondents, who had proceeded on the basis that market trading was at that time within the protection of the Town and Country Planning General Development Order; (2) because the resolution, passed after only seven days of trading in 1987, was to issue and serve a notice on or after the 15th day of trading, whereas the 1971 Act requires consideration of past or present, but not prospective, breaches of planning control in relation to enforcement; (3)
F because expediency was not considered and the 1971 Act requires proper consideration of expediency and this could not have been given in relation to a prospective breach; (4) because there had been no decision to issue the enforcement notice on April 8 1988, which was the day on which it was in fact issued.

It was further submitted that after November 17 1987 no further decision was taken with regard to the issue and service of an enforcement notice. On March 30, the only decision taken was to serve a stop notice, and it was not

possible at the committee meeting on April 25 1988 to ratify either the decision A
of November 17 1987 or the issue of the notice on April 8 1988, which, being based on the November decision, was likewise a nullity. Finally, it was submitted that the meetings of March 30 and April 25 1988 did not expressly purport to ratify the November 17 1987 decision, and ratification is not to be implied merely because of the resolutions on those days in relation to stop notices.

Held The application was dismissed. B

1. The only construction that can be placed on the resolution reached by the committee on November 17 1987 was that the local planning authority believed, albeit wrongly, that there was at that time no breach of planning control. It is for the respondent local planning authority to show, in the first instance, that there was, in accordance with the terms of section 87 of the Town and Country Planning Act 1971, a breach of planning control *apparent to them: Tidswell* v *Secretary of State for the Environment* (1976) 34 P & CR C
152. The decision on November 17 1987 was therefore unlawful: see p 45D.

2. An *ultra vires* act cannot be ratified. Accordingly the resolution of November 17 1987 was incapable of subsequent ratification by the committee meetings on March 30 1988 and April 25 1988: see p 45G.

3. At the committee meeting on March 30 1988 the committee resolved to serve a stop notice and were at the same time deciding that an enforcement notice should be issued on the new basis of a planning breach, namely an D
alleged permanent use of the site as a market. As a consequence, the enforcement notice issued on April 8 1988 was the valid consequence of a proper decision and the related stop notice was served with full consideration of relevant matters; it was impossible to contend that the decision to serve a further stop notice was perverse or irrational: see p 46.

It was common ground that at all material times there had been a breach of planning control by the applicant. Having regard to these matters, even if the enforcement notice and consequent stop notice were held invalid there would E
have been a disinclination to exercise any discretion in favour of granting judicial review.

Cases referred to in the judgment

Associated Provincial Picture Houses Ltd v *Wednesbury Corporation* [1948] 1 KB 223; [1947] 2 All ER 680; (1947) 45 LGR 635, CA

R v *Secretary of State for the Environment, ex parte Hillingdon London Borough* F
Council [1986] 1 WLR 192; [1986] 1 All ER 810

R v *Tonbridge and Malling Borough Council, ex parte Waller* (1987) unreported

Tidswell v *Secretary of State for the Environment* (1976) 34 P & CR 152; 241 EG 83; [1977] JPL 104, DC

Westminster City Council v *Jones* (1981) 80 LGR 241; [1981] JPL 750

Application for judicial review

This was an application for judicial review, by way of *certiorari,* to quash an G
enforcement notice and a stop notice issued by the respondent local planning authority under the Town and Country Planning Act 1971.

Richard Gordon (instructed by Dawson Mason & Carr) appeared for the applicant, Mrs Wendy Hobday.

Stephen Hockman (instructed by the solicitor to the Rochester upon Medway City Council) appeared for the respondent.

The following judgment was delivered. H

ROSE J: In this matter, the applicant, pursuant to leave granted by Nolan J on April 29 1988, seeks judicial review, in particular by way of *certiorari,* to quash an enforcement notice and a stop notice issued by the respondent local planning authority, under the Town and Country Planning Act 1971, on April 8 and 26 1988 respectively. The arguments before me have been confined to the validity of those two notices.

The relevant history is as follows. The applicant is a partner in Wendy Fairs Market Operators. On December 23 1986 she and her husband bought land which was the former site of the old Invicta Cinema in the High Street in Chatham. They intended to use it as a market if, as is alleged, planning permission were granted. An application for planning permission for such use had indeed been submitted a few days before they completed the purchase.

There followed discussions and correspondence between the applicant and the respondents. The upshot was that the application for planning permission was delayed and, on December 7 1987, the applicant entered an appeal under section 36 of the 1971 Act. That appeal was ultimately dismissed on September 6 1988.

In the meantime, on November 27 1987, the applicant had learnt, in the first instance by reading a local newspaper, that the respondents had, on November 17, decided to issue an enforcement notice and stop notice in respect of her use of the land as a market. She had started so to use the land on October 31 1987.

The decision on November 17, and the circumstances in which it came about, appears from pp 157 and 158 of the bundle. At p 157 are the planning officer's notes. They include this passage:

> . . . a letter has been received from the agent acting, stating that trading has now commenced in accordance with the provisions of Class IV(2) of the Schedule to the Town and Country Planning General Development Order — this allows the land to be used for the holding of a market for not more than 14 days in any calendar year.
>
> Site is being kept under observation; trading started on Saturday 31st October and appears to be taking place on Saturdays, Wednesdays and Thursdays.
>
> Additional Recommendation:
>
> Enforcement proceedings are commenced and a "Stop Notice" be served to take effect immediately after the 14-day period has expired.

At p 158, it appears that it was resolved:

> . . . concerning Plan ME/87/3, on or after the 15th day of the use of the site as an open-air market, the appropriate Enforcement Notice be issued and served and at the same time, Stop Notices prohibiting the breach of planning control referred to in the Enforcement Notice also be served . . .

On March 30 1988 the respondents' development services committee, which had before them a report prepared on March 22 1987, resolved to serve a stop notice. The material part of that report is at pp 163 and 164, and is in these terms:

> 46. Enforcement and stop notices — Wendy Fairs Market, High Street, Chatham . . .
>
> Members are reminded that at Minute 1767(B) of the 17th November 1987, authorisation was given for enforcement action to be taken in respect of the unauthorised use of the above land for holding markets.
>
> The matter is complicated by virtue of the provisions of Class IV of the General Development Order 1977 which permits the temporary use of land for the holding of markets for 14 days in any calendar year. Wendy Fairs (the market organisers) have been claiming the benefit of these development rights for the purpose of holding markets before their application for this use had been formally considered.
>
> Some further legal research has been undertaken regarding this matter. The result of this research is a finding that the permitted development rights referred to above only apply to temporary uses. The authority for this is the case of *Tidswell* v *Secretary of State for the Environment and Thurrock Borough Council* (1976) 34 P & CR 152; 241 EG 83. It was held that where the

> evidence suggested that the use was not temporary but was a permanent use A
> the benefit of the General Development Order was not available.
>
> It is considered that the facts for the purposes of the Wendy Fairs application, clearly show a desire and intention for a permanent market. In outline these facts consist of the application for the permanent use of the site as a market, the negotiations with Officers regarding the access to the site, the negotiations with Officers regarding the proposed development itself and its alleged suitability at the proposed site, the appeal against non-determination B
> of the application and the clear intention of the market operators to establish a permanent market on the site evidenced from advertisements and correspondence with Officers.
>
> In view of the above it is considered that an Enforcement Notice served at the present time would be technically valid. At the time Members authorised enforcement action they also considered and authorised the service of a Stop Notice. Members are reminded that under certain conditions, compensation is payable where a Stop Notice has been served. The condition which would C
> be relevant for this purpose is where, on appeal, it is decided that the Enforcement Notice was invalid. For the reasons set out above, it is not considered that there would be such a finding in the present case.
>
> However before serving a Stop Notice it is considered appropriate to draw Members' attention to the circumstances in which it would be served i.e. upon reliance of an Enforcement Notice alleging a permanent use of the site for the holding of a market, and to obtain Members' further authorisation to serve a D
> Stop Notice.

It then appears from the minutes of the meeting of March 30, which are at pp 165 and 166, as follows:

> 3072 Enforcement and stop notices — Wendy Fairs Market, High Street, Chatham
>
> The solicitor reminded Members that at Minute 1767(B) of the 17th November 1987, authorisation had been given for Enforcement Action to be E
> taken in respect of the unauthorised use of land at High Street, Chatham, for holding Markets. He advised Members of the problems and the legal research which he had undertaken. He drew Members' attention to the circumstances in which a Stop Notice would be served and requested their further authorisation for its service.
>
> Resolved as a matter of urgency: — that a Stop Notice be served.

The enforcement notice and a stop notice were, in consequence, sent to the F
applicant on April 8 1988, a day on which the site had only been used for less than 14 days in 1988.

On April 15 1988 the applicant sought and obtained from Hodgson J leave to apply for judicial review of that first stop notice, following which the respondents withdrew it but purported to serve a second stop notice dated April 14 1988. Through her solicitors, the applicant objected to that second stop notice, substantially on the grounds that the first one had been objected G
to, and the respondents agreed to suspend its operation.

On April 25 1988, the respondents resolved to serve a third notice, which is the one about which complaint is presently made, and to withdraw the second stop notice.

The circumstances in which that came about are that there was before the committee, on April 25, and indeed circulated beforehand with the agenda, a H
report which is set out in the bundle, starting at p 112. I will read certain passages in it:

> 1. This report informs members of the current position with regard to enforcement action . . . and seeks a decision as to the future course of conduct to be followed.

There is then, at para 6, a reference to what had transpired on November 17 1987, though there is a gloss applied in these terms:

> A In view of the possibility that the temporary use of the site as a market might be permitted development, it was recommended that enforcement proceedings should be commenced and a Stop Notice served . . .

Para 11 refers to the meeting on March 30 1988 at which:

> Members were advised that in the light of further legal research, it appeared that the benefit of the General Development Order would not be available to
> B the Wendy Fairs Market.

The report then went on in para 17 to refer to counsel's advice having been taken, and at para 18(a)(i) to that part of the advice which said that the council were entitled to consider the question of whether it was expedient to serve a stop notice, at the same time as they considered whether to issue and serve an enforcement notice. There was a reference to the irrelevance of charter rights
C made by counsel. At para 18(c), the members of the committee were specifically asked to take into account that no discussions had taken place with regard to the service of a stop notice, notwithstanding the guidance given by the Department of the Environment in Circular 4/87. At (e), members were referred to the timetable of events and asked to consider whether the planning objections were sufficiently serious to justify the service of a further stop notice. They were likewise asked to consider that matter in the context of
D planning objections referred to at (f). At (g), there appears this passage:

> The frames upon which the market stalls are erected are permanent fixtures in the sense that they have been buried in the ground and the surface of the ground has been recovered with tarmac.

There is a reference to counsel in consequence having advised that the council were entitled to conclude that there had been a breach of planning
E control as at that date.

The committee, at p 139, as appears from the minutes, were therefore specifically reminded of certain matters in that report, including the provisions of Circular 4/87 and the timetable of events. The members resolved that the second stop notice, to which I have referred, be withdrawn, and that the solicitor be instructed to serve the third stop notice, which is the subject of attack. The present application, as I have indicated, refers to that stop notice
F and to the enforcement notice issued on April 8.

There is also, as I have indicated, an appeal pending to the Secretary of State under section 88 of the 1971 Act against the enforcement notice.

Against that historical background, it is the submission of Mr Gordon, on behalf of the applicant, that both the enforcement notice and the stop notice are invalid and should be quashed. It is convenient, before turning to the details of that submission, to set out the legal framework within which it is
G made.

Section 87(1) of the Act provides as follows:

> Where it appears to the local planning authority that there has been a breach of planning control after the end of 1963, then, subject to the following provisions of this section, the authority, if they consider it expedient to do so having regard to the provisions of the development plan and to any other material considerations, may issue a notice requiring the breach to be
> H remedied and serve copies of the notice in accordance with subsection (5) of this section.

Section 88 gives the right of appeal to the Secretary of State against an enforcement notice, to which I have already referred. Section 88A gives the Secretary of State power to quash the notice and to correct defects and errors in it. Section 89 provides penalties for failure to comply with an enforcement notice.

With regard to stop notices, section 90(1) provides: A

> Where in respect of any land the local planning authority—
> (*a*) have served a copy of an enforcement notice requiring a breach of planning control to be remedied; but
> (*b*) consider it expedient to prevent, before the expiry of the period allowed for compliance with the notice, the carrying out of any activity which is, or is included in, a matter alleged by the notice to constitute the breach, B
> then . . . they may at any time before the notice takes effect serve a notice . . . prohibiting the carrying out of that activity on the land . . .

There is in the statute no right of appeal against a stop notice.

It is common ground that before an enforcement notice can be issued and a stop notice served, a decision must be made by the planning authority. (See *per* Woolf J, as he then was, in *R* v *Secretary of State for the Environment, ex parte Hillingdon London Borough Council* [1986] 1 WLR 192 at p 194H.) C

It is also common ground that the validity or otherwise of the enforcement notice and stop notice in the present case turns on the meetings and decisions of the respondents' development services committee on November 17 1987, March 30 1988 and April 25 1988.

With regard to the enforcement notice, Mr Gordon, in an attractive argument, submits that in the light of section 87(1), the decision of November D
17 1987 was a nullity outside the respondents' statutory powers: first, because there was no breach apparent to the respondents, who proceeded on the basis that trading was at that time within the protection of the General Development Order for temporary trading; second, because the resolution passed after only seven days of trading in 1987 was to issue and serve a notice on or after the 15th day of trading, whereas the Act requires past or present but not prospective breach; third, because expediency was not considered and it requires proper E
consideration, which could not be given in relation to a prospective breach; fourth, because it was not a decision to issue an enforcement notice on April 8, which was the day on which it was in fact issued.

Mr Gordon further submits that after November 17, no further decision was taken with regard to the issue and service of an enforcement notice. On March 30, he says, the only decision taken was to serve a stop notice, and it was not possible at the meeting on April 25 to ratify either the decision of F
November 17 or the issue of the notice on April 8 which, being based on the November 17 decision, was likewise a nullity. He cited the decision of an inspector referred to in [1977] JPL 604.

Furthermore, he submitted, the meetings of March 30 and April 25 did not expressly purport to ratify the November 17 decision, and ratification is not to be implied merely because of the resolutions on those days in relation to stop G
notices.

With regard to the stop notice, it is common ground that if the enforcement notice fails, so must the stop notice. Mr Gordon submits that even if the enforcement notice is valid, the stop notice fails on *Wednesbury* rationality grounds. He relies on the contents of Circular 4/87, Annex 2 of which reflects the intention of the legislature that a stop notice should be used only when the breach is so serious that the local planning authority are justified in not H
waiting for the enforcement notice to take effect and that local planning authorities should ensure that there is available a quick and thorough assessment of the likely consequences of serving such a notice. There should be a precise cost benefit assessment and discussion should take place with the person affected before serving a stop notice.

Mr Gordon submits that the cost benefit analysis referred to at pp 136 and 137 of the bundle was inadequate. I read that which there appears:

> Cost benefit assessment of the Stop Notice
>
> The costs arising from the serving of the Stop Notice will fall on the appellant and will arise from the loss of rent from the traders. In addition there will be a loss of revenue to the individual traders. The Local Authority is not in a position to assess these costs.
>
> The benefits accruing from the service of the Stop Notice will be the improvement to the visual amenity of the neighbourhood. These will occur primarily on trading days . . . and will arise from the removal of untidy objects scattered over the site, including the access and spreading on to the highway. Litter has been a particular problem. Complaints have been received and the City Housing and Environment Manager comments that extra manpower has been required on Saturday evenings to clear litter from the High Street. The problem has varied, but was particularly acute on those days when market trading was brisk.
>
> A further benefit will be the removal of a potential traffic hazard. The Local Authority is very concerned about the movement of vehicles into and out of the site onto the High Street.

There is a reference to the substantial movement of pedestrians and to the risk of accident. There is this conclusion:

> It is difficult to state who would benefit most. Certainly local traders, some of whom have complained about litter, will benefit, together with pedestrians and shoppers using the High Street.

Mr Gordon further submits that there were no discussions with the applicant, and there was no proper regard to the planning appeal in accordance with the approach which he suggests Nolan J adopted when granting leave. It is clear, submits Mr Gordon, that the respondents wanted to issue a stop notice from November 1987 before any consideration was given to the questions of litter and safety that subsequently appeared.

For the respondents, with regard to the enforcement notice, Mr Hockman submits that at all material times there was an apparent breach of planning control. Use of the site, mostly on Saturdays, but also on Wednesdays and Thursdays, listed at pp 167 and 168 of the bundle, started on October 31 1987 and has continued up to the present, amounting to 14 days in 1987, 63 days in 1988, and a steadily increasing number of days this year.

He relies on *Tidswell* v *Secretary of State for the Environment* (1976) 34 P & CR 152, in particular at pp 155 and 156, as establishing that there is no duty cast on a planning authority to investigate, before issuing an enforcement notice, whether a developer can bring himself within some exemption or permission. It is sufficient if development appears to have taken place without planning permission. He submits that the applicant has failed to show an absence of belief in development in the local authority and that the test provided by section 87 is an objective one. On November 17, the only possible factual conclusion was, he submits, that there was an apparent breach of planning control and, in resolving as they did, the planning authority were merely exercising their discretion not to issue an enforcement notice until after 14 days. He submits that in order to succeed in her attack on the November 17 decision, the applicant must be able to point to some acceptance of lawful use by the local authority.

He further submits that, even if there were no apparent breach on November 17, this did not make the decision on that date unlawful. Section 87 permits consideration to be given at any time provided the conditions as to breach and expediency in that section are fulfilled on issue of the notice. Whereas section 87 relates to the issue of an enforcement notice, section 90 deals with the service of a stop notice after service of an enforcement notice. There can therefore, perfectly properly, he submits, be resolutions on the

A same occasion to serve both. He refers to a passage in the judgment of Whitford J in *Westminster City Council* v *Jones* reported in *The Times* on June 11 1981.*

Expediency, submits Mr Hockman, can be considered at any time. At the March 30 meeting, consideration was not limited to the stop notice. The enforcement notice was specifically referred to, both in the report at p 163 and in the minute at p 165. The report specifically referred to *Tidswell*, and the advice given was that there was then a permanent market outside the B permission in the General Development Order for temporary use. Accordingly, at the time when the enforcement notice was issued by an official on April 8, there was an actual and apparent breach of planning control.

As I understand him, Mr Hockman further submits that the decision of November 17 and the issue of the enforcement notice on April 8 were, by adoption, ratified by the decisions on March 30 and April 25. C

He relies on an unreported decision of Schiemann J in *R* v *Tonbridge and Malling Borough Council, ex parte Waller*, November 6 1987, at pp 13 to 16 of the transcript. He submits that such ratification clothed the original acts with validity.

As to the stop notice, Mr Hockman relies on the detailed analysis in the report before the meeting on April 25. The guidelines as to expediency and cost benefit analysis were, he says, properly followed and, in particular, further D discussions could have served no useful purpose. The *Wednesbury* attack on the stop notice therefore, he submits, fails.

With regard to the meeting on November 17 1987, I am unable to accept Mr Hockman's submissions. In my judgment, the only construction to be placed on the resolution on that day, in the light of the recommendation which preceded it, is that the local authority believed, albeit wrongly, that there was at that time no breach of planning control. No other construction is possible, E as it seems to me, of the decision to issue and serve an enforcement notice on or after the 15th day of use. There was no consideration then given to the point made in the *Tidswell* case, that use, albeit for only 5, 7, 9 or 14 days, might be of a permanent character rather than merely temporary. It is, in my judgment, for the respondents to show, in the first instance, that there was, in accordance with the terms of section 87, a breach of planning control *apparent to them*.

F On November 17, this was plainly not so. It appeared to them that there was then no breach, but that there would be on day 15 of the use. Put the other way, they proceeded on the basis that there was then a lawful use. In my judgment, the *Tidswell* case does not assist them.

I accept the validity of the distinction which Mr Hockman draws between issue of an enforcement notice in section 87 and service of a stop notice in section 90, but this, as it seems to me, does not avail the respondents with G regard to the November 17 decision, for the reason which I have given.

Could it be said that the decision of November 17 could have been ratified by the meetings of March 30 or April 25? In my judgment, it could not, and there is nothing in the judgment of Schiemann J to suggest that it could. On the contrary, as appears from p 13 of the transcript of that judgment, it was common ground before him that an *ultra vires* act could not have been ratified. In my judgment, the resolution of November 17 was not one which H the respondents had power to make; it was *ultra vires;* it was incapable of subsequent ratification. I am unable to accept that the issue of the notice by an officer on April 8 was capable of ratification in the absence of a valid decision by the local planning authority to issue that notice.

Accordingly, so far as the enforcement notice is concerned, this leaves for

* (1981) 80 LGR 241; [1981] JPL 750.

consideration the meeting of March 30. In this connection, I am unable to accept Mr Gordon's contention that the only decision taken on March 30 was to serve a stop notice. The report and the minute both specifically refer to enforcement notices. The report discusses *Tidswell,* advises that there is permanent use of the site as a market, refers to the planning appeal, and concludes that an enforcement notice "served at the present time" would be valid.

It is true that the report proceeds on the assumption, false as I have found it to be, that valid authorisation for enforcement action had previously been given. But it concludes with a request for further authorisation of a stop notice in reliance on an enforcement notice alleging permanent use of the site as a market. This, be it said, was, in my judgment, a correct but different basis from that of the decision of November 17.

It is clear from the minute that this report was specifically brought to the committee's attention. This being so, I have no hesitation in concluding that in resolving to serve a stop notice, the committee were, at the same time, deciding that an enforcement notice should be issued on the new basis of planning breach and, in doing so, did not improperly fail to take expediency into account.

I should add, for the sake of completeness, that, though Mr Gordon made a passing reference to the possible improper influence of charter rights on the decision to serve an enforcement notice, no reference to these appears in the documents relating to the meetings on November 17 or March 30, and the only reference on April 25 was in the context of counsel's advice, advising that they would be an irrelevant consideration. There is no evidence before me that charter rights played any part, improper or otherwise, in the decisions that I have to consider. Accordingly, the enforcement notice served on April 8 was, in my judgment, the valid consequence of a proper decision.

With regard to the stop notice, it seems to me that the detailed contents of the report, starting at p 112, and parts of which I have read, and which was submitted to the committee with the agenda, and the specific consideration given, as the minutes reveal, to Circular 4/87, make it quite impossible to contend that the decision to serve a further stop notice was perverse or irrational in the *Wednesbury* sense. This challenge, likewise, fails.

I should add this. The relief here sought is discretionary. A statutory procedure for appeal against enforcement notices is provided and is indeed being pursued by this applicant. It is likely to be heard very soon. Furthermore, it is common ground that, at all material times, there has been a breach of planning control by this applicant. Having regard to these matters, even if I had taken the view that the enforcement notice and consequently the stop notice were invalid, I would have been disinclined to exercise my discretion in favour of granting judicial review.

For all of these reasons, this application fails.

Application dismissed with costs. No stay of execution granted.

Care Link and the Trustees of the Formby Settlement *v* Secretary of State for the Environment and another

QUEEN'S BENCH DIVISION

Mr Malcolm Spence QC (sitting as a deputy judge of the Queen's Bench Division)

February 15 1989

Appeal against deemed refusal for development adjoining listed building — Inspector recommending grant of planning permission and listed building consent — Secretary of State dismissing appeal for part but granting planning permission for restoration and conversion of listed building — Whether A
Secretary of State gave proper consideration to the need for all the development to finance work to listed building — Whether Secretary of State entitled to grant planning permission for only part of the development sought in the application

The applicants proposed to restore Formby Hall, Merseyside, a listed B
building. They had planned to finance the cost of this restoration by constructing an institution for the care of elderly and physically handicapped people, consisting of 24 residential units, two nursing homes, two residential-care homes, central facilities, and a community centre, and also incorporating the restoration and conversion into flats of Formby Hall. The inspector who held the inquiry into the deemed refusal to grant the necessary planning permission and listed building consent by the second respondents, Sefton
Metropolitan Borough Council, recommended that planning permission and C
listed building consent be granted. However, the Secretary of State for the Environment did not agree and dismissed the appeal in so far as it related to the 24 residential units, the two nursing homes, the two residential-care homes, the central facilities and the community centre. He granted planning permission and listed building consent for the restoration and conversion of Formby Hall into eight flats. The applicants challenged those decisions on two
grounds: (1) that the Secretary of State had failed to pay regard to the fact that D
the whole object of the exercise was to finance the cost of the restoration of Formby Hall out of proceeds of the proposed institution as a whole; he had not faced up to the consequences properly of what he was proposing to do, as by granting the restricted planning permission for the eight flats, the result of dismissing the appeal in relation to the rest of the proposed development would be that the Hall would, in effect, fall down; and (2) that the Secretary of
State was not entitled to grant planning permission for part only of the E
application by virtue of the principle in *Bernard Wheatcroft Ltd* v *Secretary of State for the Environment* (1980) 43 P&CR 233.

Held 1. The application was allowed as, although the Secretary of State recognised that the scheme was the last opportunity to preserve and restore the listed building, Formby Hall, when granting planning permission in the
abbreviated way that he had, he had not appreciated the financial F
consequence of what he had done. There was no indication that he had formed a proper view in the light of the unambiguous conclusions of the inspector in favour of granting planning permission: see pp 50H-51B.

2. It always has been lawful to grant planning permission for part only of that for which an application is made; see section 38(2) and (4) of the Town and Country Planning Act 1971. It was not contended in the present case that the application was severable in the way that it was severable in *Kent County Council* v *Secretary of State for the Environment* (1976) 33 P&CR 70. In the

A present case the application was for an institution which embraced all the features which follow thereafter in the definition; it was a single application for a single institution. The principle that governs the present case is that in the *Wheatcroft* case, namely whether there was a substantial difference between the application and the development permitted by the planning permission such that the development is so changed that to grant it would be to deprive those who should have been consulted on the changed development of the
B opportunity of such consultation. There will in practice be few cases where it would be unlawful to permit less than that which has been applied for. This is because where what is proposed to be permitted falls within the ambit of the application, be it however small a part of the application, those who should have been consulted will already have objected, either at the inquiry or by letter, to the whole scheme, and their views will already be known. In the present case, in relation to the *Wheatcroft* principle, there was no obligation
C on the Secretary of State to have canvassed everyone again: see pp 51F-53. The application would not have been granted in relation to this point had it stood by itself.

Cases referred to in the judgment

Bernard Wheatcroft Ltd v *Secretary of State for the Environment* (1980) 43 P&CR 233; 257 EG 934; [1982] JPL 37, CA

Kent County Council v *Secretary of State for the Environment* (1976) 75 LGR 452; 33 P&CR 70; [1977] EGD 843; 241 EG 83; [1976] JPL 755

D *Wessex Regional Health Authority* v *Salisbury District Council* [1984] JPL 344

Application under section 246 of the Town and Country Planning Act 1971

This was an application under section 246 of the Town and Country Planning Act 1971 to challenge a decision of the Secretary of State for the Environment in relation to his decision to, in part, dismiss and, in part, allow an appeal against a deemed refusal of planning permission and listed building consent by the second
E respondents, Sefton Metropolitan Borough Council.

David Woolley QC and Peter Harrison (instructed by Macfarlanes) appeared for the applicants.

Duncan Ouseley (instructed by the Treasury Solicitor) appeared for the first respondent, the Secretary of State for the Environment.

Charles Cross (instructed by the solicitor to Sefton Metropolitan Borough Council) appeared for the second respondents.

F *The following judgment was delivered.*

MR MALCOLM SPENCE QC: Formby Hall in Merseyside is a building of considerable architectural and historic interest. It is in a very poor state of repair and structural condition. Indeed, it is only just capable of repair, and then only if the repairs are carried out at once. The applicants proposed to restore it and also the dovecote which is in the same grounds. In order to
G finance the cost of this restoration they propose to construct an institution for the care of elderly and physically handicapped people, consisting of 24 residential units, two nursing homes and two residential-care homes, central facilities, a community centre and incorporating the restoration and conversion into flats of Formby Hall.

They made an application for planning permission to this end. They also
H made application for listed building consent for the restoration and conversion into eight flats of Formby Hall and for the restoration of the dovecote.

The local planning authority did not determine these applications and the applicants appealed to the Secretary of State for the Environment. An inquiry was held. The inspector recommended that planning permission and listed building consent be granted as applied for. However, the Secretary of State did not agree with his inspector and he dismissed the appeal in so far as it related to the 24 residential units, the two nursing homes, the two residential-care homes, the central facilities and the community centre. However, he granted planning permission for the restoration and conversion of Formby Hall into eight flats, the restoration of the dovecote, and he granted listed building consent as applied for.

The applicants appeal to this court to quash the Secretary of State's decision wherein he refused permission for the 24 residential units, the two nursing homes, the two residential-care homes, the central facilities and the community centre. They say that he failed to pay regard to the fact that the whole object of the exercise was to finance the cost of the restoration out of the proceeds of the institution as a whole. In the inspector's report at para 7, they are recorded as saying: ". . . their restoration is entirely dependent on the grant of planning permission for the continuing care institution".

The inspector made some important findings of fact, two of which I shall refer to here: "(ii) the Hall is in a poor state of repair and suffers from a number of structural defects;" and "(iv) its restoration is likely to cost between £800,000 and £985,000 excluding professional fees".

The inspector accepted the applicants' case in every respect. In para 60 of his report, he said:

> Formby Hall is of architectural and historical interest and is well worth restoring. Its structural condition and state of repair are such that it is just capable of being restored. However, unless action is taken soon — in weeks rather than years — the building will be beyond redemption. The appeal proposals appear to offer one last chance of saving Formby Hall. Listed building consent should be granted.

Para 61 states:

> However, granting listed building consent will not alone ensure the rescue of Formby Hall. The restoration works are part of a package which also contains proposals for substantial new buildings in the Green Belt. In my opinion, developments which would in other circumstances be inappropriate to the Green Belt should not be permitted solely because they would enable the financing of the restoration of a listed building. The new development should either be appropriate to the Green Belt, for example, an institution standing in extensive grounds, or there should be very special circumstances justifying an exception to the restrictive Green Belt policy.

There then followed a paragraph in which he dealt with the question as to whether what was proposed was an institution in extensive grounds, upon which he held favourably to the applicants.

However, the Secretary of State at para 5 of his decision letter said:

> The Secretary of State accepts the Inspector's findings of fact. However, while he notes that the proposed development may be seen to contain institutional elements, he is not persuaded that the proposals as a whole can be regarded as an "institution standing in extensive grounds". He therefore concludes that the proposals do not fall within the categories of development appropriate in a Green Belt, as listed in paragraph 13 of Planning Policy Guidance Note 2. Moreover, while the Secretary of State notes that the appeal proposals appear to offer a last opportunity to preserve and restore a Grade II* listed building, he does not accept that in this case this constitutes the very special circumstances necessary to justify development involving substantial new buildings in the Green Belt. He therefore concludes that the proposed development would represent a major infraction of Green Belt policy. For these reasons, therefore, the Secretary of State dismisses your appeal as it relates to the construction of 24 Residential Units, 2 Nursing Homes, 2 Residential Care Homes, central facilities and community centre.

He then went on in para 6 to say: "However the Secretary of State notes what has been said about the deteriorating condition of Formby Hall and agrees that everything possible should be done to preserve and restore the building". He then granted the planning permission which I have already described.

Mr Woolley, who appears on behalf of the applicants, did not wish to question the Secretary of State's finding that it would not be an "institution standing in extensive grounds" because he conceded that that was a matter of planning judgment for the Secretary of State. Mr Woolley also accepts that the Secretary of State did not leave entirely out of account the finances. That appears from the sentence which I have read from para 5 of the decision letter, where he said: ". . . the Secretary of State notes that the appeal proposals appear to offer a last opportunity to preserve and restore a Grade II* listed building".

However, he attacks the Secretary of State's decision on two main grounds. First of all, Mr Woolley submits that it would have to be acknowledged by the Secretary of State that the result of dismissing this appeal would be that the Hall would, in effect, fall down. Having said what he did in para 5, he did not face up to the consequences properly of what he was proposing to do, namely, to grant this restricted planning permission for eight flats, being, as it were, a mere block of flats that could be occupied by anyone be they elderly or not. Mr Woolley said that this was a highly material consideration and that the grant of the permission which he gave showed that he either misunderstood the nature of the proposals or misunderstood the financial consequences of his decision. Mr Ouseley appears for the first respondent, the Secretary of State, and Mr Cross appears for the second respondents, the local planning authority, the Sefton Metropolitan Borough Council. On this point, Mr Cross adopted all of Mr Ouseley's submissions. Mr Ouseley said that it is impossible to imagine that the Secretary of State, having said what he did in para 5, in particular the sentence which I have just reread as to this being a last opportunity to preserve and restore a listed building, would have forgotten what he said when he got to para 6 or misunderstood it. He said that what the Secretary of State did was to consider the listed building application, which was an application which was before him in any event. He observed that there was no objection to it and, therefore, it obviously had to be granted. The application for listed building consent indicated that the works thereunder were to be works for conversion into eight flats. They could, of course, only be used as flats if planning permission were granted to that end. The only application that was before the Secretary of State which would enable him to grant permission for their use as flats was the application which is the subject of these proceedings. So the Secretary of State permitted the relevant part of the application.

Mr Ouseley also said that when the Secretary of State used the language which he did in para 6 — "everything possible should be done to preserve and restore the building" — he was not to be taken to have misunderstood the financial aspects of the case, but to be recognising that he was in fact granting very little by the permission which he was proposing, it being a permission that there was no ground for withholding, and that it might help, even if it was most unlikely to help, in view of what he had already accepted, if he were to grant it. Mr Ouseley said that that was all the Secretary of State meant by those words.

In my judgment, it would not be safe to let this decision stand as it is. What the Secretary of State has done on this application for an institution (it is conceded by all the parties before this court that it is an application for an institution embracing not only new buildings but also Formby Hall itself, as part of the institution) is to grant planning permission for a perfectly ordinary block of flats. In particular, although the Secretary of State has in that brief sentence in the middle of para 5 recognised that the scheme was the last

opportunity to preserve and restore the building, in my view the Secretary of A
State, when granting the permission in the abbreviated way that he has in para
6, has not, to use the language of Mr Woolley, appreciated the full horror of
what he was doing.

If the Secretary of State had concluded the letter at the end of para 5 with a
complete refusal of planning permission, it might have been another matter,
but by going on in para 6 to grant a perfectly useless planning permission,
doubt is cast, in the mind, upon the approach by the Secretary of State to the B
financial aspects of this case. In my judgment, there should be further
opportunity for representations to be made to the Secretary of State in order
for him to perform a proper view on this extremely important aspect. As I say,
as the letter stands at the moment there is no indication that he has formed a
proper view upon it, particularly in the light of the unambiguous conclusions
of the inspector. C

Second, Mr Woolley submitted that the Secretary of State was not entitled
to permit part only of the application by virtue of the principle in *Bernard
Wheatcroft Ltd* v *Secretary of State for the Environment* (1980) 43 P&CR 233.
On this aspect, it is important that I should mention at the outset that neither
Mr Woolley nor Mr Ouseley contended that the application was severable in
the way that it was severable in the earlier case of *Kent County Council* v
Secretary of State for the Environment (1976) 33 P&CR 70. D

Mr Cross, who appeared on behalf of the planning authority, drew my
attention in particular to p 76 of the *Kent County Council* case where Sir
Douglas Frank QC (sitting as a deputy judge of the High Court) said: "It
further seems to me that, as a matter of common sense, the determining
authority can grant as much of the development applied for as they think
should be permitted". Then I need not read on.

However, in my view, the principle in the *Kent County Council* case does E
not apply in the present case because, unlike the circumstances in the *Kent
County Council* case, in this case the application is not severable. As accepted
by both Mr Woolley and Mr Ouseley on looking carefully at the form of the
application, it is an application for an institution which embraces all the
features which follow thereafter in the definition, which I have already read. It
is a single application for a single institution. Mr Woolley emphasised that,
anyway, the whole project was a total package. Nothing that I say in this case F
affects anything that was said in the decision in the *Kent County Council* case
to the effect that it is lawful to permit part only of an application and refuse the
other part.

In my judgment, it is and always has been lawful to grant permission for
part only of that for which application is made. That is entirely clear from
section 38(2) and (4) of the 1971 Act, which is the present equivalent of sections
which have been in the legislation almost since the outset. The *Kent County* G
Council case is a good example, but in practice there must be many more
examples where permission can properly be granted in respect of part only of
the land to which this application for permission related or where conditions
may properly be imposed so as, in effect, to permit less than that which is
applied for. In these cases no further consultation is required at all. In my
judgment, this principle will govern most such cases. However, it is agreed H
that it does not govern the present case for the reason I have just mentioned
and that the principle which does govern the present case is the principle in the
Wheatcroft case. At p 241, Forbes J said:

> I conclude, for my part, that there is no principle of law that prevents the Secretary of State from imposing conditions that have the effect of reducing the permitted development below the development applied for except where the application is severable. The Secretary of State clearly directed himself

> that there was such a principle and thus fell into error, and his decision must be quashed.
>
> I should add a rider. The true test is, I feel sure, that accepted by both counsel: is the effect of the conditional planning permission to allow development that is in substance not that which was applied for? Of course, in deciding whether or not there is a substantial difference the local planning authority or the Secretary of State will be exercising a judgment, and a judgment with which the courts will not ordinarily interfere unless it is manifestly unreasonably exercised. The main, but not the only, criterion on which that judgment should be exercised is whether the development is so changed that to grant it would be to deprive those who should have been consulted on the changed development of the opportunity of such consultation, and I use these words to cover all the matters of this kind with which Part III of the Act of 1971 deals.

The judgment of Glidewell J, as he then was, in *Wessex Regional Health Authority* v *Salisbury District Council and Secretary of State for the Environment* [1984] JPL 344, is also helpful. In that case, prior to the public inquiry, the appellants had submitted a further layout plan showing 37 houses rather than the 48 which had been shown earlier. The inspector, in his decision letter, stated that the appellants had invited him to consider the argument that if he thought 48 houses were too many then 37 houses were acceptable, he would grant planning permission subject to a condition limiting the number to 37 according to the decision in *Wheatcroft*. Glidewell J went on to say that it was for the Secretary of State or his inspector to determine whether the difference was substantial. The report stated at p 346:

> With some reluctance he (Glidewell J) agreed with [counsel for the respondents]. As a matter of law, he did not see how he could properly say that the inspector had not acted within the powers of the Act and he certainly did not feel able to say that his decision was so unreasonable that no reasonable inspector could have arrived at it. He emphasised that he arrived at this decision with some reluctance.

I am bound to say that I also would have reached that conclusion with reluctance.

There will in practice be few cases where it would be unlawful to permit less than that which has been applied for, for where what is proposed to be permitted falls within the ambit of the applicaton, be it however small a part of the application, because those who should have been consulted — I am looking in particular at the language of Forbes J in *Wheatcroft* — will already have objected, either at the inquiry or by letter, to the whole scheme, and their views will already be known. Thus, it will make little difference whether they were consulted again or not.

If one takes the circumstances of this very case — I have to assume for the purposes of this part of the judgment that the Secretary of State had not fallen foul of the proper standards in the manner to which I have already adverted — if he had demonstrated that he had considered the financial aspects properly but had nevertheless granted the permission which he has granted in his decision letter, then any further representations, first of all, from the applicants would have been of little or no avail. Likewise, the local planning authority had never objected to flats in the Hall. Likewise, there would have been nothing else that the third parties would have wanted to say. That is all perfectly obvious from the report.

Indeed, Mr. Cross, on behalf of the local planning authority, has said, so far as his clients, the local planning authority, are concerned, that they have not been prejudiced by the course which the Secretary of State has adopted.

Accordingly, I do not accept that for this reason, namely, that related to the *Wheatcroft* principle, the Secretary of State should have canvassed everyone again. That would have achieved nothing.

Furthermore, if this point had stood by itself, I should not have been inclined to exercise the court's discretion to quash his decision on this ground, for the same reason that nothing would have been achieved by canvassing the views of the parties again.

However, for the reason which I gave in relation to Mr Woolley's first submission, the decision will be quashed and the application allowed.

Application allowed with costs against the first respondent, the Secretary of State for the Environment.

R *v* Ogwr Borough Council and another, *ex parte* Carter Commercial Developments Ltd

QUEEN'S BENCH DIVISION
Macpherson J
February 17 1989

**Out-of-town retail schemes — Rival developments — Consent recommended for land owned by council — Land subject to commoners' rights — Judicial review sought by disappointed rival applicant — Whether rights of commoners properly considered — Whether direction on departure procedure
A properly followed — Whether alternative sites a relevant consideration — Whether applications must be considered together — Whether council's financial interest made procedure unfair**

During 1987 and 1988 13 applications for planning permission were submitted to the respondent borough council in respect of out-of-town retail schemes. Two applications were made by the applicant company to develop a retail park
B on the site of a disused hospital. One of these applications was refused in January 1988, and the other application led to an appeal which, at the date of the instant case, was before the Secretary of State for the Environment awaiting his decision. That scheme was known as the Pines Centre development. Another application was made by the Bridgend Golf Club Ltd, the second respondents, for a retail, leisure and golf course development on a site, known as Sarn Park, of which the club held a lease from the respondent
C council. That site was north of the M4, near exit 36. Coite Walia Common forms a large part of the Sarn site and is the subject of commoners' rights. The planning officer to the respondent borough council prepared a report recommending the advantages of the Sarn Park site, in particular its location adjacent to the M4 motorway. It was acknowledged that all the applications were in respect of development which would amount to a departure from the local development plan.

D In May 1988 the respondent council resolved to grant the necessary outline planning permission for the Bridgend Golf Club's application, and, following the conclusion of the recommended agreements, an outline planning permission was granted in August 1988.

The applicants moved for judicial review of the council's decision that planning permission should be granted for the golf club scheme on the grounds that: (1) the council failed to consider material considerations,
E namely that the site at Sarn Park is subject to commoners' rights and that the council were materially misinformed about the state of the negotiations with the commoners; (2) in relation to the Town and Country Planning (Development Plans) (Wales) Direction 1981, certain objections made by the applicants had not been properly put to the council in compliance with that direction; (3) that the council had failed to consider alternative sites in the light of objections to the Sarn Park site; (4) that the council had failed to consider a joint application for planning permission made on May 20 1988 by the
F applicants and the Land Authority for Wales; (5) that the council had failed to take particular and express note of the fact that they were departing from the development plan; and (6) that because the council had a financial interest as the freehold owners of Sarn Park, there must be scrupulous attention to all the steps taken and there must be no risk that there had been any unfairness in the process of reaching the decision that was ultimately reached. The decision to allow development of Sarn Park had prejudiced other schemes the subject of appeals to the Secretary of State.

Held The application was dismissed. A

1. The existence of commoners' rights cannot prevent the grant of planning permission. However, the respondents were informed of the existence of the rights and were not misled as to the state of the negotiations with the representatives of the commoners' association as to the release of those rights: see pp 59H-60B.

2. Where the Development Plans Direction 1981 is followed, and objections are received by a local planning authority following the necessary B newspaper advertisements, those objections which are truly fresh objections, or which raised materially new matters, should as a matter of common sense be put before the council before they make their decision. However, in the present case the second of two objections made on behalf of the applicants contained distinctions without any real difference and it did not set out anything which could have been ignored by the council in the first place, presented as they were with all the facts in their officials' reports: see p 60. C

3. This case was not one of those in which it was necessary for the respondents to be fully warned that they must themselves independently consider alternative sites: cf *Trusthouse Forte Hotels Ltd* v *Secretary of State for the Environment* (1986) 53 P & CR 293. The Sarn Park site had no such individually clear objections in planning terms that it was necessary for the council to look elsewhere and make the comparison referred to in the cases. The council had reached their own informed planning judgment based on comparison of the competing applications in relation to this type of scheme in D the Bridgend area: see p 61D.

4. There was no obligation on the council to consider the joint planning application made on May 20 1988, but that application really added very little, if anything, by way of opposition to the Sarn Park application. Although a local planning authority can reconsider their decision to grant planning permission up to the moment when it is signed and issued in writing, this does not mean that an authority must as a matter of law or practice open up their E formal decision upon submission of any fresh and possible competing application, right up to the last moment before grant: see p 62B.

5. Where a local planning authority are financially interested, as this one was, there must be scrupulous attention to all the steps taken and there must be no risk that there has been any unfairness in the process of reaching the decision which was ultimately reached. There was no unfairness in the conduct of the council and indeed the matter was considered very fully in the early and F later stages in accordance with the principles of law and practice: see p 63B.

Cases referred to in the judgment

Greater London Council v *Secretary of State for the Environment* (1985) 52 P&CR 158; [1986] JPL 193

R v *West Oxfordshire District Council, ex parte C H Pearce Homes Ltd* [1986] JPL 523

R v *Yeovil Borough Council, ex parte Elim Pentecostal Church Trustees* (1971) 70 LGR 142; 23 P&CR 39, DC G

Trusthouse Forte Hotels Ltd v *Secretary of State for the Environment* (1986) 53 P&CR 293; [1986] 2 EGLR 185; 279 EG 680; [1986] JPL 834

Ynystawe, Ynysforgan and Glais Gypsy Site Action Group v *Secretary of State for Wales* [1981] JPL 874

The following cases were cited in relation to the order for costs

R v *Registrar of Companies, ex parte Central Bank of India* [1986] 1 QB 1114; [1986] 2 WLR 177; [1986] 1 All ER 105, CA H

Vale of Glamorgan Borough Council v *Secretary of State for Wales* (1986) 52 P&CR 418; [1986] JPL 198

Application for judicial review

This was an application for judicial review of the decision of the first respondent, Ogwr Borough Council, to grant planning permission to the second respondent, Bridgend Golf Club Ltd.

A *Nicholas Nardecchia (instructed by Freshfields) appeared for the applicants.*
Geoffrey Stephenson (instructed by the solicitor to the Ogwr Borough Council) appeared for the first respondent.
Glanville Pugh QC and Charles Parsley (instructed by Simon North & Nam, of Cardiff) appeared for the second respondents, the Bridgend Golf Club Ltd.

The following judgment was delivered.

B **MACPHERSON J:** In 1987 development was much in the air in Bridgend, in Mid Glamorgan. A whole series of applications was made connected with out-of-town retailing. I believe that there were 13 applications. Of particular significance were four of these applications.

On October 16 1987, Carter Commercial Developments Ltd applied on behalf of the Dunraven Family Trustees to develop a retail park adjoining and including the former Cefn Hirgoed Hospital. C That application was, in fact, refused on January 27 1988. On October 16 1987 another application was made by the same applicants which was never determined. That application led to an appeal dated December 11 1987 against this non-determination. The appeal was heard on and after August 2 1988. The Secretary of State's decision as to that appeal has yet to be given. That development is known in this case as the Pines Centre Development.

D The Land Authority for Wales applied at about the same time for a major development at Heol West Plas and Brackla (see p 140 of Bundle B). There was also made, in September 1987, the application which led to the outline planning permission with which I am concerned. That was an application, 87/1094, made by the Bridgend Golf Club Ltd. That proposed development was north of the M4, near exit 36, and it is known in this case as the Sarn Park application.

E It is of significance that on November 19 1987, the Mid Glamorgan County Council asked the Welsh Office to call in all the major applications for retailing and associated development around Bridgend, indicating that the grant of planning permission to any of them would amount to a departure from the local development plan. A copy of the county planning officer's reports in this connection went with that letter.

On December 3 1987 the Ogwr Borough Council planning officer reported F to his forward planning committee (see pp 55 to 63 of Bundle B). It is plain that by then the official preference was for a large development adjacent to the M4 and near junction 36. A development with the potential to "fulfil a truly sub-regional/regional role" was envisaged (see, in particular, p 61 of Bundle B).

On January 4 1988, the Welsh Office declined to call in the relevant planning applications, indicating that it believed that Ogwr Borough Council G were "fully alive to the need to take into account the factors listed in the letter of November 19 and to give them proper weight". The Welsh Office noted that there was likely to be a departure from the development plan whatever happened and said that this would be borne in mind if any such reference should be made to the department.

The ball was therefore back in Ogwr's court and the borough planning officer proceeded to marry together the various applications. He produced H then and later very full and comprehensive reports on each application. Not surprisingly, and in my judgment perfectly properly, he set out again his view that "the ideal location for such development is adjacent to the M4 motorway and junction 36 at Sarn".

But he submitted for consideration maps and information concerning all the relevant applications. In each instance he set out details of the application and a comprehensive summary of observations received from consultees and interested parties. Often an objector to one application was himself, or itself, an applicant. In this context I refer to the Bridgend Golf Club Ltd report, 87/1094, which starts at p 73 of Bundle B. It is of some importance to note that at p 80, the report says, in terms, that this scheme had been "advertised by the applicants in accordance with section 26 of the Town and Country Planning Act 1971 and by the Department in accordance with Welsh Office advice regarding discretionary advertising of major proposals".

This earlier advertising had evidently prompted objections and comments. These are set out in the succeeding pages. Simply for example, I refer to the objection of the commoners of Coity Walia Common, which forms a large part of the Sarn site. This includes full information as to the registration of these commoners' rights and of their objection to the scheme. It includes express reference to the previous grant by Ogwr of permission for the golf club to set up the 18-hole golf course on the site.

I also refer to the objection of Dunraven Welsh Estates (p 86) which, among other things, expressly refers to the Greater Bridgend Local Plan adopted by Ogwr Borough Council and to the recreation and tourism policy 9.3.32 as to the provision of the 18-hole golf course at Cefn Hirgoed.

It seems to be a fair comment that it is therefore odd that the present applicants before me should doubt whether Ogwr Borough Council took into account their (the council's) own departure from a local plan, since they themselves drew the council's attention to that very fact. In any event, it seems to me most unlikely that all involved were otherwise than perfectly well aware of that departure, as is shown by reference to departure in the letter to the Welsh Office in November 1987.

The Pines Centre applications, 87/1221 and 87/1222, are dealt with at p 67 and following. They received similar full treatment, but unfortunately for Dunraven the official's recommendation was that these applications should be refused. In contrast, of course, is the recommendation properly made in accordance with practice by the officials that the Sarn application should be granted subject, as can be seen in the reports, to all necessary agreements and conditions.

A main reason for recommending the rejection of the Pines Centre application, 87/1221, was that it was not such as to "create a regional focal point in which the wider concepts in the fields of leisure, recreation and tourism, both indoor and outdoor, are combined with the proposed retailing facilities in an integrated package". There was also at that stage a traffic objection.

At p 115 there is information about the second Pines Centre application, which had not received a decision in due time and was, of course, already subject to an appeal so that no decision could be given upon it by Ogwr in January 1988. In due course, on January 27 1988, the first Pines Centre application failed, leaving the present applicants to their appeal route.

On February 29 1988, yet another Dunraven application was launched, but that one remains undetermined and unappealed. It has, so I am told, now the benefit of a Highways Act agreement to nullify or mitigate the earlier traffic objection to the Pines Centre site. This comes about apparently because of a transfer of that agreement to the February application from yet another application, which we have known as the "joint application", made by the present applicants and the Land Authority for Wales, who had, of course, also earlier failed on their own.

I will refer again to that joint application. It was not made until May 20 1988. It was withdrawn on September 16 1988 and, by means which are not referred to in evidence but which were related to me by counsel, the available Highways Act agreement, which had attached to that defunct joint application, now attaches to the Dunraven application of February 28. In due course the Sarn Park application made its successful way through all the planning processes. I note, of course, as Mr Nardecchia asked me to do, the changes in the actual terms of the intended or proposed permission on the way through, although I am bound to say that I do not find such adjustment or change surprising. In this connection the relevant references are p 226 (February 11 1988), p 231 (May 5 1988), and p 233, together with its appendix at p 237 (July 21 1988).

On May 5 1988 the council resolved as follows:

> That subject to the terms of the necessary legal agreements as amended by 1 above, being reported back to the Sub-Committee for approval prior to their execution, the Borough Planning Officer be authorised to issue the outline planning permission subject to the conditions previously agreed by the Sub-Committee and as amended by 2 and 3 above.

The relevant agreements were the section 52 agreement regarding phasing and so on, and a Highways Act agreement. These were in due course completed on July 22 and July 29 1988 respectively. They had earlier been approved (see pp 264 to 283 of Bundle B). On August 5 1988 outline permission was signed in favour of Bridgend Golf Club Ltd subject to full conditions (see p 284).

I turn to the present application in which Carter Commercial Developments Ltd apply to quash the permission granted to Bridgend Golf Club Ltd. The case is to some extent a strange one, since Carter/Dunraven have an undecided appeal outstanding. At the hearing of that appeal it was said that the grant of leave in respect of this present application went some way *prima facie* to indicate that the court might say that the council had acted improperly and had been activated in part by financial considerations. This should never have been said. Particularly is this so when in this court there has been no attempt to impugn the good faith of Ogwr Borough Council in connection with any profit which may accrue to them as landlords of the Sarn site.

Quite rightly, Mr Nardecchia has said no such thing. He put the case in this regard simply on the basis of unfairness and particularly on the basis of cumulative unfairness, pointing out simply that where a council are financially involved as the owners of a site, they must both follow, and be seen to follow, the full and correct procedures and all that they are at each stage required to do. I accept that that is the duty of the borough council.

I turn, then, to the particular areas of complaint advanced here, and in doing so I bear in mind, of course, that this is judicial review. I am not assessing the merits of this or any other application. I am considering the decision-making process. If I find any error of law or any flaw in that process, then in my discretion I can strike down the planning permission. If the council have been misled or have failed to take into account relevant and material matters, as they should, or have acted unfairly, then in each case there could be material available upon which I could base an order for review.

To this extent I agree with Mr Nardecchia that error or unreasonableness upon those grounds is independent of the out-and-out absurdity and irrationality principle which was enshrined as long as 40 years ago in the *Wednesbury* case [1948] 1 KB 223 and which now dominates the working lives of many judges. If any error of law or unfairness or failure to act properly in granting this permission exists, then it should go. The fact that the real motive for seeking its removal is to advance the applicants' cause is really neither here nor there. Development is about profit, and competing developers are entitled to attack each other's grants provided the legal basis for such attack exists.

1 *Commoners' rights.* A

I have already referred to the commoners' objections to the Sarn Park development. There is no doubt that these rights, which probably concern about 160 commoners, subsist. The development cannot be carried on without removal of those rights either by buying them off or by having them erased from the register in Chancery proceedings. Such proceedings were begun in 1988 and were served on January 17 1989.

Of itself, in my judgment, the existence of the registration does not and B cannot prevent the grant of planning permission. Many grants are made in the knowledge that, for example, restrictive covenants must be removed or altered before a project can go ahead. But the matter is relevant as to viability and was taken into account in this case. The council knew perfectly well what the basic position was and that the rights still existed when they resolved to make their grant. The only question here, in my judgment, is whether the council were misled or there is a reasonable likelihood that they were misled by C what they were told about the possible agreement between the commoners and developers.

Mr Nardecchia argues that the council were materially misinformed about the state of the negotiation. For this purpose I must refer to two passages in the evidence, the first starting at p 79. This is an extract from the affidavit of Mr Harris, who is the solicitor to the commoners. He says that on February 8 1988 he met the solicitor representing Bridgend Golf Club, who wanted him to write D a letter to the Ogwr Borough Council saying that he had reached agreement with the Bridgend Golf Club. Then he goes on in this way:

> I indicated that it was only I who had reached agreement on the price which I could recommend to the Commoners on the assumption that the Commoners were prepared to agree to the principle of withdrawing their objection. All the letter could say was that we were merely having discussions on the basis that E my Firm were Professional Advisers and any such terms would have to be approved by the Commoners at General Meeting. The next day, the 9th February 1988, I met Mr Daniel . . . I discussed the matter fully with Mr Daniel. I informed him that agreement had been reached between the Professional Advisers on a figure which could be recommended to the Commoners should they wish to withdraw their objection. I made it clear to Mr Daniel that there had been a Meeting of the Commoners on the 28th F January 1988 at which there had been much opposition to the proposals put forward by Bridgend Golf Club and that all that was happening at the moment was that negotiations were still going on.

As a result of that conversation, Mr Daniel reported to the committee.

I turn to p 30 of the evidence to see what the committee of the council were told on February 11 1988. The passage reads:

> Shortly before the meeting I spoke to the solicitors acting respectively for the G Commoners Association and for the applicants for planning permission, Bridgend Golf Club Limited, and was told that terms had been provisionally agreed between their clients' representatives for the withdrawal of the Commoners Association's objection to the Council's application under section 13 of the Commons Registration Act 1965. I therefore informed the meeting of the situation and outlined the provisionally agreed terms as I understood them. I explained that I had been informed that the draft terms H agreed between the representatives of the two parties concerned would be put before a general meeting of the Commoners Association in a few weeks but that until they were approved by the Commoners Association in general meeting there could be no legally binding agreement between the Association and the proposed developers.

I am wholly unable to say that this amounted to a material misdirection. The commoners' solicitors had said that the professional advisers had reached agreement on a figure which could be recommended to the commoners. No binding agreement had been or could possibly have been made until the commoners' general meeting and the council were told that. I do not believe that this account of what Mr Daniel had been told was likely, in any event, to have moved the council, since one of their members was alert to the fact that the litigation in Chancery was still alive, and provisional agreement as to terms between representatives can often be far from final agreement.

I do not accept, therefore, that the council were misled. This ground, as now set out, is in any event to some extent a gloss on the original ground which concerned the time-limits during which the development had to be completed. Since the matter will probably be heard or perhaps dealt with by agreement in the summer of 1989 I do not accept that the time argument is valid either. In any event, time can be extended if that proves to be essential. I reject this first independent ground upon which this application is based.

2 *The Town and Country Planning (Development Plans) (Wales) Direction 1981*

As Mr Nardecchia indicates, there is no contest at all but that the direction was applicable in the circumstances of this case. In any event, the council's officials recommended that the council should follow the procedure laid down by the direction and they did so. Para 3 of the direction states:

> A departure application which the local planning authority do not propose to refuse shall, before the grant of permission thereon, be advertised (unless previously advertised in pursuance of other planning requirements) by the local planning authority in a local newspaper circulating in the locality . . .

Primarily the object of the direction is to give to the Secretary of State for Wales an opportunity to restrict the grant or require a reference of the application to him before the grant might be authorised. I refer in this regard to para 5 which so states. I see the strength of Mr Nardecchia's argument that impliedly a council should be shown any objections raised if those objections are truly fresh objections or raise materially new matters. Whether as a matter of law it is strictly necessary that this should be done I am not sure and I do not rule, but as a matter of common sense I can see that in those circumstances it can be said to be desirable and right that the council should see any truly fresh objections which come in. But, in my judgment, in the present case this was simply not required.

It can, I suppose, be said that a line-by-line comparison of the objection raised in January by Dunraven to the golf club's application and the letter of March 11 1988 does show some differences, but, in my judgment, they are distinctions without any real difference. The real truth is that Carters were saying in March 1988 simply that they believed that the council had taken "insufficient account" of the uncertain viability of the Sarn Park project and of the unrealistic nature of the shopping list of leisure items and so on.

I do not believe that the letter sets out anything which could have been ignored by the council in the first place, presented as they were with all the facts in their official's comparative reports and with the objections which were set out with care in those reports. Certainly there was no reaction from the Secretary of State to these objections, which were all ordinary planning matters which must have been considered in any event. I have already indicated the correspondence which passed between the Welsh Office and the Mid Glamorgan Council indicating that the Welsh Office might in any event wish to call in these applications. Having not done so in the first instance, it is very unlikely, in my judgment, that the letter of March 11 1988 could or ever would have made any difference.

It would certainly also be hard, in my judgment, to remove this grant from the Bridgend Golf Club Ltd because the local authority did more than they were required to do and advertised in accordance with the direction when in fact they had already advertised in pursuance of other planning requirements. In my judgment, the letter of March 11 1988 did not need to be put before the council. It did not truly raise fresh or fundamental matters which the council needed to consider anew, and I reject this ground.

3 *Alternative sites*

I have been directed to the relevant cases in this regard. It is enough, in my judgment, to mention *Trusthouse Forte Hotels Ltd* v *Secretary of State for the Environment* (1986) 53 P&CR 293; *Greater London Council* v *Secretary of State for the Environment* (1985) 52 P&CR 158; and *Ynystawe, Ynysforgan and Glais Gypsy Site Action Group* v *Secretary of State for Wales and West Glamorgan County Council* [1981] JPL 874.

The law in respect of applications of the present kind is usefully summarised, in my judgment, in the *Ynystawe* case at p 876. I do not propose to read the passage which we considered in argument and which I have reconsidered before giving this judgment.

Probably it is often at inquiry level that the question of alternative sites is discussed and argued; but that it can arise at the time of a grant by a local authority I am prepared to accept. Here, however, in my judgment, the case is not one of those in which it was necessary for the council to be fully warned that they must themselves independently consider the question in the *Trusthouse* case sense. I refer in that regard to p 299 of the judgment of Simon Brown J and to his summary of the effect of the spate of cases in this respect which had gathered by 1987.

In particular I read numbered para (2):

> Where, however, there are clear planning objections to development upon a particular site then it may well be relevant and indeed necessary to consider whether there is a more appropriate alternative site elsewhere. This is particularly so when the development is bound to have significant adverse effects and where the major argument advanced in support of the application is that the need for the development outweighs the planning disadvantages inherent in it.

All the relevant sites and in particular the major ones which were considered by the Ogwr Borough Council had planning objections. This site had no such individually clear objections in planning terms that it was, in my judgment, necessary for the council to look elsewhere and make the comparison referred to in the cases. In any event, virtually every large developable site was up for grant in this round of applications and the relative appropriateness of one site over another was in this case part of the comparative process which the council had to perform in considering the 13 applications and in particular the major ones.

The council by themselves and through their officials regarded the Sarn Park site as really the only one which offered that regional role which they were seeking in respect of the new venture. Whether others disagreed with this conclusion is another matter, but it was certainly, in my judgment, a perfectly sustainable view looking at the various applications together and comparing them as the council must have done in their consideration of all the applications in January 1988.

Mr Nardecchia argues that there are no signs here that there was any guidance as to how the competing alternative schemes compared in town and country planning terms and no evaluation or comparison put forward of their respective features for the council to consider. This is a somewhat different

A point, but in any event, in my judgment, it is invalid. Submission of all the planning objections to each scheme in the various reports achieved this as far as it was required by law. What the council must do is to reach their own informed planning judgment based on comparison of the competing applications. I fail to see how it can be said that this was not achieved and I do not believe that the officials were bound further to instruct or advise the council than they did.

B

4 *Joint application*

Again, I do not accept that it was necessary for the council to have been asked to consider this application made on May 20 1988. I accept entirely the legal position as to grant as set out in *R* v *West Oxfordshire District Council, ex parte CH Pearce Homes Ltd* [1986] JPL 523 and *R* v *Yeovil Borough Council,* C *ex parte Trustees of Elim Pentecostal Church, Yeovil* (1971) 23 P&CR 39. It is established by these cases that a council can reconsider their grant up to the moment when it is signed and issued in writing. But this does not, in my judgment, mean that the council must as a matter of law or practice open up their formal decision upon submission of any fresh and possibly competing application, right up to the last moment before grant.

I suppose that the joint application did have some additional features which D might have gone towards its own success, if it had gone ahead, and I am certainly prepared to assume that that is so. But since it was truly an amalgamation of the old Land Authority for Wales and Pines Centre application, it really added very little, if anything, by way of opposition to the Sarn Park application. In planning terms there was no reason, in my judgment, why the fresh one should be put to the council to try to change their mind.

E That application was, in any event, withdrawn, so that I have had some difficulty in seeing why it would be just now to revive it and say that its contents should persuade me that the council might have decided differently if they had seen it and then go on to exercise my discretion to quash the presently existing Bridgend Golf Club Ltd permission. It seems to me that because of the decease of the joint application it would be unjust to use it as a stick with which to beat the golf club. I reject this ground.

F

5 *Departure*

I have probably dealt with this aspect of the case as much as is necessary, but the allegation is that the council failed to take particular and express note of the fact that they were departing from the development plans which required that the Sarn Park site should be used as an 18-hole golf course. The fact is, G however, that this matter was advertised as a departure. The Secretary of State was told that each application involved departure. The council must have known perfectly well that this was so, and I see no need whatsoever for further specific mention of the matter in order, so to speak, to "show their working" in respect of this aspect of the case.

6 *Unfairness*

H The question is, says Mr Nardecchia, would a reasonable man observing the whole process have said that the council acted unfairly? It is perfectly true that the council have a basic financial interest in this development, since the ownership of the land is vested in the council, but that was known to all. There is not, and there never has been, any particular matter of bias or financial gain advanced as a reason for objection in this case. That was mooted in the original grounds and, as Mr Stephenson says, the present attack, launched with moderation and skill by Mr Nardecchia, is differently slanted.

What Mr Nardecchia says, in truth, is that where a council are financially interested as this one was, there must be scrupulous attention to all the steps taken and there must be no risk that there has been any unfairness in the process of reaching the decision which was ultimately reached. That I accept. This court is always alert to try to detect any unfairness in a decision of this kind.

At the root of judicial review is natural justice and the fairness in the conduct of administrative and executive proceedings. And the question that I have to answer is whether the council have acted in such a way that they could be said not to have been able properly to exercise their discretion. Did the council genuinely and impartially exercise their discretion?

I say at once that I see no unfairness individually or cumulatively in this matter whatsoever. I have looked at each component part of Mr Nardecchia's argument, I hope with great care. I have considered each aspect of the grounds which are put forward, but I see no unfairness either in any of the individual grounds or in the consideration of any of the matters by the council which have been put before me in this case.

It does not seem to me possible to spell unfairness in the round out of the conduct of the council as evidenced by the affidavits and the documents which have been put before me. Indeed it appears to me that when the matter was considered very fully in the early and later stages the council considered everything that they were required to do with care and in accordance with the principles of law and practice which they were required to follow.

In those circumstances, the applicants before me today must await their own appeal and perhaps even yet another approach to the council on their last undecided application. I do not believe that it would be right or just to disturb the permission at present held by Bridgend Golf Club Ltd. I therefore reject this application and it must be dismissed.

Applications for costs were made on behalf of the first respondent and the Bridgend Golf Club Ltd, the second respondent, during which the following cases were cited:

R v *Registrar of Companies, ex parte Central Bank of India* [1986] 1 QB 1114
Vale of Glamorgan Borough Council v *Secretary of State for Wales* (1986) 52 P&CR 418

His lordship concluded his judgment:

In this case both respondents apply for costs against the applicants. The usual order is that one set of costs is paid in circumstances of this kind in applications for judicial review, but there are exceptions to the usual order and, in my judgment, this is one of them. In these proceedings the applicants alleged against the local authority bad faith as can be seen from the later grounds set out. That has been moderated by Mr Nardecchia on instructions in this court, but the fact is that that was a live issue until the matter was argued before me. In those circumstances, the concentration of the local authority was plainly upon that allegation, which is an unpleasant and serious one to make, whereas the Bridgend Golf Club was, in my judgment, having been served, wholly entitled to come to argue on its own account, if necessary, that the grant of planning permission should be maintained in its favour.

It is true that there has not, as it turned out, been much difference of argument from the point of view of the two respondents, but in view of the way in which the matter was launched and the risk that the local authority might well have been involved primarily in defeating an allegation of bad faith, it seems to me that this is a case which is out of the ordinary and exceptional in the terms of the direction as to costs set out in Order 53 of the Rules of the

A Supreme Court. In those circumstances, it seems to me perfectly right, proper and just that both those who have been successful should be awarded their costs. The respondents will both be paid their costs by the applicants, those costs to be on the standard basis.

Application dismissed with costs to both respondents.

Robbins *v* Secretary of State for the Environment and another

HOUSE OF LORDS
Lords Bridge of Harwich, Brandon of Oakbrook, Templeman, Ackner and Lowry
March 2 1989

Listed building — Repairs notice — Compulsory purchase — Repairs notice specifying works of preservation and works of restoration — Whether repairs notice satisfied condition precedent for compulsory purchase — Proper A
content of repairs notice — Whether repairs notice including works beyond preservation valid — Building-owner's appeal dismissed

Willesborough Windmill, Ashford, Kent, was erected in 1868, and in 1951 it was listed as a building of special architectural or historic interest. The appellant acquired it in 1969 and converted it into a dwellinghouse. The condition of the windmill deteriorated, and in October 1983 the second B
respondents, the Ashford Borough Council, served a repairs notice on the appellant pursuant to section 115 of the Town and Country Planning Act 1971. The repairs notice specified 20 items of work that the second respondents considered were, in the terms of section 115, "reasonably necessary for the proper preservation of the building". Some of the items were works of an emergency nature or were repairs to prevent further deterioration of the structure; other items, such as the reconstruction of the catwalk and fantail and work to the sweeps, were later accepted as being works of C
restoration rather than preservation, as some parts of the structure were missing from the building when it was listed in 1951 and were also missing at the date of the repairs notice in 1983.

Following the failure of the appellant to comply with the repairs notice, the second respondents made a compulsory purchase order, under section 114 of the 1971 Act, in respect of the windmill. The appellant objected to the draft order and, following a public local inquiry and the recommendations of his D
inspector, in December 1986 the first respondent, the Secretary of State for the Environment, confirmed the order, as he concluded that a repairs notice specifying works that were not "preservation" was not invalid and he was not satisfied that the appellant had taken reasonable steps for preserving the building. The appellant challenged the confirmation of that order by the first respondent on the ground that the repairs notice, the necessary condition precedent to the exercise of compulsory purchase, was invalid as it did not satisfy the requirements of section 115 of the 1971 Act. The decision of Mr E
Malcolm Pill QC (sitting as a deputy High Court judge) to dismiss the appellant's application to quash the decision of the first respondent to confirm the order was affirmed by the Court of Appeal ([1988] 2 EGLR 205). It was contended on behalf of the appellant that a repairs notice, a condition precedent of compulsory purchase, may only specify works for "proper preservation" of a listed building, that a line must be drawn between preservation and restoration and that, on the true construction of sections 114 F
and 115, works cannot be considered necessary for the proper preservation of a listed building which are not directed to the preservation of the building as it subsists as the date when the repairs notice is served.

Held The appeal was dismissed.

The word "preservation" has to be given its ordinary meaning in contrast to "restoration"; this imposes an objective limitation which must be applied in considering the works specified in the notice: see p 71C.

A The date of the listing of the building, and not the date of the repairs notice, is the relevant date for determining whether works are those of repair or restoration: see p 72F.

Where a repairs notice contains items of work that are not within the scope of "preservation", it remains a valid notice notwithstanding that it also includes invalid items. Provided that the list of valid items is sufficiently substantial to support a conclusion by the Secretary of State, in the event that
B the specified works are not carried out, that reasonable steps are not being taken for properly preserving the building, the invalid items may simply be disregarded: see p 73. The owner's remedy is to challenge a notice containing excessive items in the magistrates' court under section 114(6) of the 1971 Act either because they are bad in law or in fact: see p 74B.

No cases are referred to in the opinions

C **Appeal against a decision of the Court of Appeal**

This was an appeal against a decision of the Court of Appeal ([1988] 2 EGLR 205) dismissing an appeal against the refusal by Mr Malcolm Pill QC (sitting as a deputy High Court judge) (November 13 1987) of an application by the appellant under section 23 of the Acquisition of Land Act 1981 to challenge the decision of the first respondent, the Secretary of State for the Environment, who had confirmed a compulsory purchase order made by the second respondents, the Ashford
D Borough Council.

Michael Barnes QC and Richard Hayward (instructed by Edwin Coe & Calder Woods, for Roderick O'Driscoll & Partners, of Maidstone) appeared for the appellants.

Jeremy Sullivan QC and Mark Low (instructed by the Treasury Solicitor and by Sharpe Pritchard for the solicitor to the Ashford Borough Council) appeared for the first and second respondents.

E

The following opinions were delivered.

LORD BRIDGE OF HARWICH: My Lords, The appellant challenges a compulsory purchase order made by the Ashford Borough Council on October 15 1984 and confirmed by the Secretary of State for the Environment on December 16 1986 which authorises the council to acquire compulsorily the
F Willesborough Windmill. The authorisation was given pursuant to section 114 of the Town and Country Planning Act 1971 on the ground that it appeared to the Secretary of State that "reasonable steps are not being taken for properly preserving" the windmill, which is a listed building. It is a condition precedent to a compulsory acquisition under section 114 that at least two months before initiating the compulsory purchase proceedings the acquiring authority should serve on the owner of the building a notice under
G section 115 "specifying the works which they consider reasonably necessary for the proper preservation of the building". The notice is referred to in the section as a "repairs notice".

The ground of the appellant's challenge is that the repairs notice which was served on him by the council on October 7 1983 did not satisfy the condition precedent in that some of the works specified as considered by the council to be
H reasonably necessary for the proper preservation of the windmill were incapable, on the true construction of section 115, of being so considered and that the inclusion of these excessive items invalidated the notice.

The challenge thus raises two distinct questions of statutory construction. First, what, in the context of sections 114 and 115, is the scope of the "proper preservation" of a listed building which the works specified in a repairs notice under section 115 may be directed to achieve? Second, if the notice specifies some works falling within that scope but also others which exceed it, is the statutory condition precedent to compulsory acquisition satisfied or is the notice ineffective for that purpose?

The appellant applied to the High Court pursuant to section 23 of the Acquisition of Land Act 1981 for the quashing of the compulsory purchase order. The application was dismissed by Mr Malcolm Pill QC, sitting as a deputy judge of the Queen's Bench Division. His decision was affirmed by the Court of Appeal (Slade, Glidewell and Russell LJJ): [see [1988] 2 EGLR 205]. The appellant now appeals by leave of your Lordships' House.

For the purpose of deciding the appeal the relevant facts may be quite shortly stated. The windmill was built in 1868. It continued in operation as a windmill until 1938. In September 1951 it was included in a list of buildings of special architectural or historic interest pursuant to section 30(1) of the Town and Country Planning Act 1947. The notes entered in the list descriptive of the building read as follows:

> Built in 1868 by John Hill of Ashford, mill-wright. Rect. brick base of 2 s. Above this an octagonal smock mill of white weather-boarding with a platform and railing round above the base. Sash ws. with gl. bars intact. Hooded cap. Fantail and sweeps partly missing. The windmill is still worked as a mill but not by wind. Unusually good condition.

In 1969 the windmill was acquired by the appellant and converted to a dwellinghouse. Over the years the condition of the windmill deteriorated greatly. In particular, by the time the council served the repairs notice in October 1983 the platform and railing round the base of the mill ("the catwalk") had mostly decayed or been removed, parts of the fantail had been removed and what remained was in danger of collapse, and very little was left of the sweeps.

The repairs notice served under section 115 specified 20 items of work which the council considered reasonably necessary for the proper preservation of the building. It is now accepted that 14 of those items, those numbered 1-12, 14 and 18 in the list, specified either works of an emergency nature (dismantling the dangerous remains of the fantail) or works in the nature of repairs required to prevent further deterioration of the structure or of parts of the windmill which were subsisting at the date of service of the notice. The remaining items, however, went beyond this. They included, *inter alia*, the complete reconstruction of the catwalk and the fantail and the renewal of the stocks and whips (but not the framework or shutters) of the sweeps. The work specified for the catwalk and the sweeps was to be "to the original standard or to a standard approved by the council". I shall for convenience refer to items 13, 15, 16, 17, 19 and 20 collectively as "the restoration items".

When the compulsory purchase order was made, the appellant duly objected to its confirmation and a public inquiry was held in December 1985 by an inspector appointed by the Secretary of State. In a lengthy report, dated January 10 1986, the inspector records the submissions made to him and sets out his findings of fact. Objection was taken on behalf of the appellant that the inclusion in the repairs notice of the restoration items was unlawful. Since service of the repairs notice, the ruinous remains of the fantail have been dismantled, but apart from that it is clear that very little work had been done in response to the notice. The inspector found that:

> the steps taken can in my opinion only be described as preliminaries for the eventual preservation of the building rather than as substantial works for its proper preservation.

The inspector concluded:

> that no reasonable steps are being taken for properly preserving the building

> and that it is expedient to make provision for its preservation and to authorise compulsory acquisition for that purpose.

The crucial paragraphs of the Secretary of State's decision letter, dated December 16 1986, read, so far as presently material, as follows:

> 5. Careful consideration has been given to the legal points set out by the inspector at paragraphs 27 to 44 of his report and his opinion at paragraph 45 that they are for the Secretary of State to decide. Representations were made to the effect that: (i) some of the works included in the repairs notice were appropriate to the restoration of the building rather than to its preservation. The Secretary of State takes the view that the question whether works are properly considered to be reasonably necessary for the proper preservation of a building is bound, to a certain extent, to be one of fact and degree. He agrees with the inspector's view that in certain instances works which might normally be considered more appropriate to restoration can, in other circumstances, be regarded as necessary for the proper preservation of a building. In the Secretary of State's view, the repairs required to preserve the building contained in the council's repairs notice do not include such items as would invalidate the notice although the Secretary of State accepts that had all, or most, or even a substantial amount of the works required by items 1-12, 14 and 18 of the schedule to the notice been carried out, he would have been satisfied that reasonable steps were being taken for properly preserving the building; . . .
>
> 6. The Secretary of State, having considered the legal points raised on behalf of the owners of the mill, considers that there is no legal impediment to prevent him from reaching a decision on the compulsory purchase order on the merits of the case.
>
> 7. The inspector's findings of fact and conclusions have been carefully considered. The inspector's conclusion that the building is a particularly important one which warrants every effort being made to preserve it is accepted, as is also his conclusion that no reasonable steps are being taken for properly preserving it. . .
>
> 9. The Secretary of State accepts the inspector's findings of fact and his recommendation and he is satisfied that it is expedient to make provision for the preservation of the building and to authorise its compulsory purchase for that purpose. He has accordingly decided to confirm the order without modification.

Before examining the rival submissions of the parties it is essential to consider the general scheme of the Act of 1971 in relation to the preservation of buildings of special architectural or historic interest. Lists of such buildings are compiled or approved by the Secretary of State under section 54(1). Section 54(2) provides:

> In considering whether to include a building in a list compiled or approved under this section, the Secretary of State may take into account not only the building itself but also —
>
> . . .
>
> (*b*) the desirability of preserving, on the ground of its architectural or historic interest, any feature of the building consisting of a man-made object or structure fixed to the building or forming part of the land and comprised within the curtilage of the building.

As soon as may be after listing the local authority are notified and they in turn notify the owner of the building. A copy of the entry in the list is registered as a local land charge. Thus every subsequent owner will acquire with knowledge that the building is listed. Once a building has been listed under section 54, section 55 makes it a criminal offence to demolish the building or to alter or extend it in any manner which would affect its character as a building of special architectural or historic interest, unless permission, referred to in the Act as "listed building consent", has been granted by the local planning authority or the Secretary of State. The owner of a listed building is not consulted and has no right of objection before the building is listed. However, if he applies for listed building consent and it is not granted by the local planning authority, he has, on appeal to the Secretary of State, a full opportunity to canvass the merits of the listing and it is at that stage that the Secretary of State will have to reach a decision for or against the merits of preserving the building. If listed building consent is refused or granted subject to conditions, compensation is, in appropriate circumstances, payable to the owner pursuant to section 171. A B

The machinery, the Act provides to secure the preservation of a listed building, may be considered under three headings: first, sanctions for unlawful demolition or alteration; second, provision for preservation work to be undertaken by the local authority or the Secretary of State; third, provision for compulsory acquisition when reasonable steps are not being taken by the owner for properly preserving the building. C

The offence created by section 55(1) is punishable by a fine or imprisonment. In addition, when works have been carried out in contravention of section 55(1), sections 96 to 100 of the Act provide an elaborate procedure initiated by the service of a listed building enforcement notice, analogous to the enforcement notice procedure applicable to ordinary breaches of planning control, to secure the restoration of the listed building. In particular, by section 96(1)(*b*)(i) (as substituted by the Local Government and Planning (Amendment) Act 1981, section 1 and Schedule, para 9), a listed building enforcement notice may require such steps as may be specified to be taken "for restoring the building to its former state". One of the grounds of appeal available against such a notice under section 97(1)(*g*) (as substituted) is that "the requirements of the notice exceed what is necessary for restoring the building to its condition before the works were carried out". Non-compliance with a valid listed building enforcement notice attracts further criminal penalties under section 98 and the ultimate sanction to secure restoration of the listed building is provided by section 99 which allows the local planning authority to take the steps required to be taken by the enforcement notice and to recover the expenses of so doing from the owner. D E

Subject to certain limitations, section 101 provides that when it appears to the local authority or the Secretary of State that works are urgently necessary for the preservation of a listed building they may execute the works. Since April 1 1987 the significance of this provision, as an encouragement to the owner of a listed building to keep it in proper repair, has been strengthened by the addition of section 101A, introduced by the Housing and Planning Act 1986 (section 40 and Schedule 9, Part I, para 7), which enables the expenses of works executed under section 101 to be recovered from the owner. I mention this as a recent enhancement of the statutory machinery for the preservation of listed buildings, but I recognise, of course, that it cannot affect the construction of sections 114 and 115. F G

Turning now to the machinery for compulsory acquisition, I first set out so much of sections 114 and 115 (as amended) as seems to me material as follows:

> **114.**—(1) Where it appears to the Secretary of State, in the case of a building to which this section applies, that reasonable steps are not being taken for properly preserving it, the Secretary of State may authorise the council of the county or county district in which the building is situated . . . to acquire compulsorily under this section the building and any land comprising or contiguous or adjacent to it which appears to the Secretary of State to be required for preserving the building or its amenities, or for affording access to it, or for its proper control or management. H
>
> . . .

A (3) This section applies to any listed building. . . .

(4) The Secretary of State shall not make or confirm a compulsory purchase order for the acquisition of any building by virtue of this section unless he is satisfied that it is expedient to make provision for the preservation of the building and to authorise its compulsory acquisition for that purpose.

(5) The Acquisition of Land Act 1981 shall apply to the compulsory acquisition of land under this section.

B (6) Any person having an interest in a building which it is proposed to acquire compulsorily under this section may, within 28 days after the service of the notice required by section 12 of the Acquisition of Land Act 1981, apply to a magistrates' court acting for the petty sessions area within which the building is situated for an order staying further proceedings on the compulsory purchase order; and, if the court is satisfied that reasonable steps have been taken for properly preserving the building, the court shall make an order accordingly.

C (7) Any person aggrieved by the decision of a magistrates' court on an application under subsection (6) of this section may appeal against the decision to the Crown Court.

115.—(1) The compulsory purchase of a building under section 114 of this Act shall not be started by a council . . . unless at least two months previously they have served on the owner of the building, and not withdrawn, a notice under this section (in this section referred to as a "repairs notice") —

D (*a*) specifying the works which they consider reasonably necessary for the proper preservation of the building; and

(*b*) explaining the effect of sections 114 and 117 of this Act.

. . .

(4) For the purposes of this section a compulsory acquisition is started when the council . . . serve the notice required by section 12 of the Acquisition of Land Act 1981.

E Section 116 provides that the measure of compensation on compulsory acquisition of a listed building, unless compensation has already become payable under section 171 upon refusal of listed building consent or grant of such consent subject to conditions, shall be assessed on the assumption that listed building consent for demolition, alteration or extension of the building would be granted. This ensures that the open market value upon which compensation is assessed is not to be depreciated by the restriction inherent in
F the listing. Section 117 provides for a reduced level of compensation in a case where the building has been deliberately allowed to fall into disrepair for the purpose of justifying its demolition. It is not suggested that this is such a case.

Mr Michael Barnes QC, who has presented the case for the appellant with conspicuous skill and cogency, submits that a line must be drawn between preservation and restoration and that, on the true construction of sections 114
G and 115, works cannot be considered necessary for the proper preservation of a listed building which are not directed to the preservation of the building as it subsists as at the date when the repairs notice is served. He accepts, of course, that the concept of preservation in these sections cannot be limited to keeping the building in the exact condition in which it is when the notice is served, since this would frustrate the whole procedure. But he submits that no more can ever be considered necessary for preservation than such repairs as are required
H to secure whatever remains of the building from further deterioration. Thus, for example, if part of the roof is missing or a wall has become unstable, it is accepted that works to repair the roof or stabilise the wall would properly be directed to preservation of the building. But, on the other hand, if some distinct part of the building, some decorative feature of the building without structural significance or some free-standing object included in the listing under section 54(2)(*b*) has been accidentally destroyed, works to repair or replace those items would exceed the ambit of preservation and could not

properly be included in a repairs notice. This construction, Mr Barnes A
submits, accords with the ordinary meaning of the word "preservation" in
contrast with the word "restoration" found in the provisions relating to the
enforcement notice procedure in sections 96 and 97. On the basis of this
construction the restoration items specified in the repairs notice in this case
were beyond the scope of section 115.

Mr Jeremy Sullivan QC, for the respondents, submits that the phrase
"reasonably necessary for the proper preservation of the building" must be B
construed as a whole and establishes the criterion to be applied by the
acquiring authority, when they serve the repairs notice, and the Secretary of
State, when deciding whether to confirm the compulsory purchase order in the
circumstances of any particular case, which is partly a matter of fact and
degree and partly a matter of discretion. Their decisions in the application of
that criterion, he submits, are only susceptible of review by the courts on the
ground of irrationality. I readily accept, as did Mr Barnes in argument, that C
the use of the words "reasonably" and "proper" in the phrase under
consideration call for value judgments, weighing such matters as the cost and
the benefit of works required for the preservation of a listed building, and that
such judgments are entrusted to the acquiring authority under section 115 and
to the Secretary of State under section 114. But I think that the word
"preservation" has to be given its ordinary meaning in contrast with D
"restoration" and that this does impose an objective limitation which must be
applied in considering what the works specified in a repairs notice may be
directed to achieve.

The more difficult question is whether "preservation" of the listed building
in these sections refers to the preservation of the building as it was when listed
or of the building as it is when the repairs notice is served. I think the language
of the sections is capable of either construction. Indeed, both counsel adopted E
the proposition that the sections refer to preservation of the building as it
subsisted at the date of listing as an alternative to their primary submissions,
albeit with very different emphasis. Mr Barnes' bottom line, if I may use that
colloquialism, is that the preservation envisaged by sections 114 and 115 must
at least exclude the restoration of features of a building which ceased to exist
before the building was listed. Mr Sullivan's bottom line is that it must at least
include anything designed to preserve the building as it subsisted when first F
listed and that this would include, in the case of a building in disrepair at the
time of listing, works designed to remedy the disrepair and secure the building
against further deterioration.

I accept that the legislature cannot have intended that immediately
following the listing of a building it should be liable to compulsory purchase
on the ground that steps were not being taken for properly preserving it G
because the owner was unwilling to restore features of the building which had
ceased to exist before listing. I accept, on the other hand, that if what I will call
the date of listing construction is to be preferred to the date of notice
construction, a building in disrepair when listed may be the subject of a repairs
notice under section 114 specifying works necessary to prevent further
deterioration. The line between repair and restoration may not be an easy one
to draw with precision, but in practice I doubt if any great difficulty will be H
found in saying whether any particular works fall on one side of the line or the
other. The important issue is whether the date of listing or the date of notice
construction is correct.

Mr Barnes naturally places much emphasis on the specific references to
restoration which can be required to be undertaken by or at the expense of the
owner under the listed building enforcement notice procedure in sections 96 to
100. He also contends, I think rightly, that the works urgently necessary for

the preservation of a listed building, which may be undertaken by the local authority or the Secretary of State under section 101, and the expenses of which since April 1 1987 may be recovered from the owner under section 101A, can only refer to works urgently necessary to preserve the building as it subsists when the works are undertaken. These indications, he submits, point strongly in favour of the date of notice construction. I am conscious of the force of Mr Barnes' submission but I cannot accept that the considerations on which he relies are decisive. An enforcement notice under section 96 is both penal and coercive; it compels the owner to restore the building or to bear the cost of restoration. A repairs notice is in no way either penal or coercive; it is a procedural preliminary to compulsory acquisition designed to give the owner the opportunity, if he chooses, to undertake the works reasonably necessary for the proper preservation of the building as an alternative to selling it at its market value to the acquiring authority. Again, it is clear that works can only be urgently necessary for the preservation of a building, so as to justify works being undertaken pursuant to section 101, if they are necessary for the preservation of the building as it subsists when the works are carried out. Thus there is no ambiguity in section 101. I do not think it follows that the proper preservation of a listed building, which is the objective of the compulsory acquisition procedure under sections 114 and 115, is limited in the same way.

I believe that the question, whether the date of listing or the date of notice construction is correct, is to be resolved purposively by considering the underlying policy of the legislation. The public interest in the preservation of buildings of special architectural or historic interest needs no emphasis. Once a building has been listed, that public interest has been declared. If the owner seeks and is denied unconditional listed building consent he will recover any compensation payable under section 171. If a listed building falls into disrepair and that disrepair becomes apparent before the building or part of it collapses, the character of the building can be preserved, if necessary, by emergency works under section 101. But if part of a building collapses without warning or is destroyed by fire or storm damage, the character of the building as a building of special architectural or historic interest can be preserved only if the damage is made good. If the date of notice construction is correct, the compulsory purchase machinery is ineffective to serve the public interest in such cases and sections 114 and 115 are of very limited utility. On the other hand, if the date of listing construction is correct, compulsory purchase is available in such cases as the only means, if the owner is unwilling to make good the damage, of preserving the character of the building from which its special architectural or historic interest derives. I have no hesitation in concluding that the date of listing construction is to be preferred. Sections 114 and 115, given this more generous construction, do no more than enable the building as listed to be acquired and preserved at the public expense. The interest of the owner, if he is unwilling to undertake the necessary works, in retaining his property has to yield to the public interest in the same way and on the same terms as the interest of any other property owner whose property is acquired for some necessary public purpose.

I have felt it appropriate to examine this issue of construction at some length, since it was fully argued and is of obvious importance. At the end of the day, however, I do not think the result of the appeal turns on it. I accept Mr Barnes' submission that even on the date of listing construction the repairs notice served by the council specified works exceeding what could be considered reasonably necessary for the proper preservation of the windmill. The notes entered in the list in 1951 recorded that the fantail and sweeps were then partly missing. The restoration items in the repairs notice specified works for the full restoration of the fantail and partial restoration of the sweeps. The former clearly are, and the latter, in the absence of evidence as to which parts A of the sweeps were missing in 1951, must be taken to be, works for the restoration of the building to the condition that it was in before it was listed.

The Secretary of State, however, clearly did not rely on the fact that none of the works specified in relation to the restoration items had been carried out in reaching his conclusion that reasonable steps were not being taken for preserving the windmill. There was ample material on which he could reach that conclusion in reliance on the failure to carry out works specified as B reasonably necessary for the preservation of the windmill in relation to the other 14 items listed in the repairs notice. His confirmation of the compulsory purchase order, therefore, was perfectly lawful under section 114 unless, as Mr Barnes contends, the inclusion in the repairs notice of the restoration items relating to the fantail and the sweeps invalidated the remainder of the notice so that the condition precedent to compulsory purchase under section 115 was C not satisfied.

Before turning to that issue, I should observe that the statutory code relating to the preservation of listed buildings, on which I have based the opinion expressed as to the proper construction of sections 114 and 115, is the consolidating Act of 1971. A large part of that code, including the provisions now found in sections 114 and 115, was first enacted by the Town and Country Planning Act 1968 amending the original code under the Town and Country D Planning Act 1947. I leave open for future consideration, if the question should ever arise, whether works specified for the restoration of some feature of a building first listed under section 30 of the Act of 1947 which had disappeared before the Act of 1968 came into force could properly be regarded as necessary for the preservation of the listed building within the meaning of sections 114 and 115. The point does not arise here and no argument was addressed to it. It is very unlikely, I would think, to arise in the future and I E express no opinion on it.

The remaining and crucial question on which the appeal depends is whether a repairs notice which specifies a list of works considered reasonably necessary for the proper preservation of a listed building which includes a number of items within the scope of the section ("valid items") and also a number of items within the scope of the section ("invalid items") effectively satisfies the condition precedent to compulsory purchase imposed by section 115. This is a F pure question of the construction of the section and I do not think that any assistance is to be derived from authorities on severance in very different legal contexts. On the face of the language of section 115 a notice listing a number of valid items is a notice "specifying the works which they consider reasonably necessary for the proper preservation of the building" notwithstanding that it also includes invalid items. Provided that the list of valid items is sufficiently G substantial to support a conclusion by the Secretary of State, in the event that the specified works are not carried out, that reasonable steps are not being taken for properly preserving the building, it is difficult to see why the invalid items should not simply be disregarded. Mr Barnes submits, however, that the purpose of a repairs notice is to give the owner of a listed building the opportunity to avoid compulsory purchase by undertaking the works which are reasonably necessary for the proper preservation of the building and that a H notice which is excessive puts him in a dilemma; he does not know whether to carry out all the specified works or to omit those works which he considers to be excessive at the risk of having his property acquired if he is held to have been wrong. Such a dilemma may arise whenever the owner of a listed building wishes to challenge any items of works specified in the repairs notice as excessive. The ground of a challenge may be that the disputed items are excessive in fact, ie that valid items exceed what is in fact reasonably necessary for the proper preservation of the building, or it may be that the disputed items are excessive in law, ie that they are invalid items. Mr Barnes rightly concedes that a notice containing valid items which are excessive in fact satisfies the condition precedent to compulsory purchase imposed by section 115. It is otherwise, he submits, if invalid items are included because the acquiring authority, in serving the notice, has erred in law.

The dilemma of the owner is the same whether he wishes to challenge certain items in the repairs notice as being excessive in fact or excessive in law. It seems to me that recourse to the magistrates' court under section 114(6) is tailor-made to provide a solution to the dilemma in either case. This unusual provision empowers the magistrates' court to override the opinion of the acquiring authority and to pre-empt the decision of the Secretary of State in determining what works are reasonably necessary for the proper preservation of the listed building.

The procedure will operate in the following way. The owner, who wishes to retain his property in the listed building, will put in hand the works specified in the repairs notice which he admits to be necessary for its proper preservation, but not the works specified in the items listed which he wishes to dispute. On receipt of notice under section 12 of the Act of 1981 initiating the compulsory purchase proceedings, he will then apply to the magistrates' court under section 114 for an order staying those proceedings. If the magistrates' court is satisfied that he has taken reasonable steps for properly preserving the listed building by the works he has already put in hand and that the disputed items are excessive, the owner will be entitled to an order staying further proceedings on the compulsory purchase order. In relation to the disputed items, the proceedings initiated in the magistrates' court will in due course resolve all issues of both fact and law and can, if necessary, be taken on appeal to the Crown Court or on a point of law to the High Court. If at the conclusion of the proceedings it is held against the owner that some of the disputed items are reasonably necessary for the proper preservation of the listed building, he will then be able to put in hand the works specified in relation to those items, and it is inconceivable that the acquiring authority would proceed with the acquisition. If they were to do so, the owner could make a fresh application to the magistrates' court under section 114(6) which would be bound to succeed.

Being satisfied that this procedure is available to protect the owner of a listed building, who is willing to carry out such works as are reasonably necessary for its proper preservation, from any prejudice by the inclusion in a repairs notice of invalid items, I am equally satisfied that the inclusion of such items does not invalidate the remainder of the notice.

I would accordingly dismiss this appeal.

LORD BRANDON OF OAKBROOK: My Lords, For the reasons given in the speeches of my noble and learned friends, Lord Bridge of Harwich and Lord Ackner, I would dismiss the appeal.

LORD TEMPLEMAN: My Lords, For the reasons given by my noble and learned friend, Lord Bridge of Harwich, I would dismiss this appeal.

LORD ACKNER: My Lords, I have had the advantage of reading in draft the speech of my noble and learned friend Lord Bridge of Harwich. I gratefully adopt and therefore need not repeat his recital of the facts which led to this appeal.

The central issue in this appeal is the meaning of the phrase "works . . . reasonably necessary for the proper preservation" of a listed building, in section 115(1)(*a*) of the Town and Country Planning Act 1971.

The purpose of "listing" buildings and how this is achieved A
Before considering the rival contentions of the parties, the provisions of the
Act of 1971, in so far as they relate to listed buildings, must first be considered.
The purpose of "listing" buildings, is to ensure the protection and
enhancement of the local heritage of buildings. The Secretary of State,
pursuant to his statutory obligation under section 54 of the Act, compiles lists
of buildings of special architectural or historic interest or approves such lists
compiled by the Historic Buildings and Monuments Commission for England B
and Wales, or by other persons or bodies of persons as appear to him
appropriate as having special knowledge of, or interest in, buildings of
architectural or historic interest.

Under section 54(2), in considering whether to include a building in a list
compiled or approved under the section, the Secretary of State may take into
account not only the building itself, but also any respect in which its exterior C
contributes to the architectural or historic interest of any group of buildings of
which it forms part, and the desirability of preserving, on grounds of its
architectural or historic interest, any feature of the building. Moreover,
section 54(9) contains an extended definition of "listed building" to include
any object or structure fixed to the building and any object or structure within
the curtilage of the building which, although not fixed to the building, forms
part of the land and has done so since before July 1 1948. D

The purpose of "listing" buildings, to which I have referred earlier, is
achieved in the first place by making it a criminal offence for a person to
execute (section 55(1)):

> any works for the demolition of a listed building or for its alteration or
> extension in any manner which would affect its character as a building of
> special architectural or historic interest.

Quite apart from the penalties that may be incurred in carrying out E
unauthorised work to a listed building, power is given to the local authority to
issue a notice (a building enforcement notice) to restore the building to its
former state or to the state in which it would have been if the terms and
conditions of any listed building consent which had been granted had been
complied with (see section 96). If such a notice is not complied with not only
does the owner commit a further offence (see section 98), but provision is
made for the local authority to carry out the necessary work and recover the F
expenses reasonably incurred (see section 99).

Those sections relate to acts of commission. To cater for acts of *omission*,
section 101 makes provision where urgent works are necessary to preserve the
listed building. In such a case the owner is given notice in writing of the
intention to carry out the works, and under section 101A the expense of
executing those works is recovered by the authority who carry them out. G

Where the works required to preserve the building are not urgently
necessary, a notice under section 115 (referred to as a "repairs notice") may be
served by a local authority "specifying the works which they consider
reasonably necessary for the proper preservation of the building". Such a
notice, unlike a notice under the Housing Acts, does not require the *execution*
of any work and places no statutory obligation upon the owner to carry out
any work. It is a warning shot, a preliminary to compulsory acquisition of that H
listed building if it is not complied with. The notice has to explain that the local
authority considers that reasonable steps are not being taken for the proper
preserving of the building and that the works which they specify in the notice
are reasonably necessary for its proper preservation. It warns the owner that if
the works are not carried out within two months from service of the notice, it
may exercise its powers under section 114 of the Act to begin proceedings for
the compulsory purchase of the building. The notice also has to point out that

if (which was not the case in respect of this property) it appears that the building has been deliberately allowed to fall into disrepair, for the purpose of justifying its demolition and the development or redevelopment of the site, a compulsory purchase order may include a direction under section 117 for "minimum compensation"; the effect of which will be to limit the compensation otherwise payable on compulsory acquisition by requiring it to be assessed on the assumption that neither planning permission nor listed building consent under the Act would be granted for any works except to restore the building to a proper state of repair and to maintain it in such a state. In cases where the building has not been deliberately left derelict, compensation is based on the assumption that listed building consent would be granted for any works for the alteration or extension of the building or for its demolition, other than works in respect of which such consent has been applied for before the date of the order and refused or granted subject to conditions.

If the repairs notice is not complied with, and the Secretary of State is satisfied that reasonable steps are not being taken properly to preserve the building, he will authorise the local authority to acquire the building compulsorily under section 114. However, he may not make or confirm a compulsory purchase order under this section unless he is satisfied that it is expedient to make provision for the preservation of the building and to authorise its compulsory acquisition for that purpose.

Thus, it will be seen that although the owner may disregard the repairs notice, he does so at his peril, since thereby he renders the building liable to be compulsorily acquired. Thus, given that works are reasonably necessary for the proper preservation of the building, their execution is achieved either at the owner's expense (aided by such grants as may be available) or at the public expense after the public has compulsorily acquired the building. Thus, Parliament's purpose to ensure the protection and enhancement of the local heritage of buildings is fulfilled.

The construction of the words "works . . . reasonably necessary for the proper preservation of the building"

"Works" within the meaning of section 115(1)(*a*) must be works:

1. for the preservation of the building and
2. which the local authority consider reasonably necessary for its proper preservation.

There is no dispute that the works specified in the repairs notice prepared by the second respondents satisfied 2 above. The essential issue is — whether they were works for the "proper preservation of the building". This raises two questions, namely:

(i) the correct construction, in its context, of the word "preservation" and
(ii) by reference to what date is the condition of the building to be preserved — and in particular is the condition in which it is to be preserved to be that which existed as at the date of the repairs notice *or* the date of its listing?

Mr Michael Barnes QC, who has argued the case for the applicant with very great skill, has submitted that the works of preservation are confined to works necessary to prevent deterioration or further deterioration of the building from the condition in which it was at the time of the notice. Accordingly, features of the building, which on account of their architectural or historic interest had occasioned the listing of the building but which before the service of the repairs notice had been destroyed or seriously damaged, cannot properly be the subject of the repairs notice. Its preservation in its condition as at the date of the notice is all that can be required. According to this submission, it would matter not whether it was the owner's omissions or an act

of nature or a combination of both which had brought about the destruction A
or damage.

Mr Barnes' argument can be well illustrated by a particular feature of this
windmill. In the brief description in the list dated September 24 1951 there is a
specific reference to "a platform and railing round above the base," that is,
the rectangular brick base of two storeys. This is the "catwalk" and is clearly
shown in the photograph of the mill taken not long before the listing. As at the B
date of the repairs notice (October 7 1983) the catwalk was in a very poor
condition, with very little if any of the handrail intact. It would follow from
Mr Barnes' submission that although, no doubt, some work would be
necessary to prevent further deterioration of the catwalk there would be no
need to replace any of the handrail, clearly a feature at the time the building
was listed. Indeed, Mr Barnes readily accepted that it would follow, if his
submissions were valid, that if important features of this or any building were C
blown or burnt down the day before the notice was served, the repairs notice
could not extend to such features.

In my judgment, Mr Barnes' submissions would clearly produce results
contrary to the underlying policy of the Act, which is to preserve the building,
ie to keep it and its features as they existed at the date of listing in sound
condition. D

Did the inclusion in the repairs notice of works which went beyond what was
reasonably necessary for the proper preservation of the windmill render the
notice invalid?

Mr Barnes accepted that if the notice included works which went beyond what
was reasonably necessary for the proper preservation of the building, because
the notice required an unreasonable amount of preservation work to be
carried out, this did not render the notice invalid. For instance, where a roof E
was in a state of disrepair because there were a dozen or more tiles missing, to
require the roof to be completely stripped and retiled would clearly be
excessive. However, he submitted that where the requirement was excessive,
because works specified were not works of preservation at all but were works
involving restoring the building to a condition it did not enjoy at the material
date, be it the date of listing or the date of the repairs notice, whichever be the
correct date, the notice became an invalid notice, albeit that it contained a F
substantial amount of work reasonably necessary for the proper preservation
of the building. This was so because, so he submitted, the excessive demands in
the notice were not due to the incorrect assessment of the amount of
preservation work which was needed, but to a wrong construction by the
authority of the meaning of the word "preservation."

I am prepared to accept that certain of the works specified in the repairs
notice went beyond works reasonably necessary for the preservation of the G
building in the state in which it was when listed since they were works of
restoration to a pre-existing state. In the notes in the list there is a reference to
"fantail and sweeps partly missing". In the repairs notice the applicant was
required to *renew* the fantail and the sweeps. Mr Barnes submits that that is an
indication that the respondents wrongly construed the word "preserve" as
including restoration to a pre-existing state. For the purpose of his argument I H
am prepared to accept that he may well be right, particularly when one has
regard to the fact that work which was described as "long-term restoration
work" in the report of the engineers and millwrights retained by the council
was in part incorporated in the repairs notice.

I am unable, however, to understand why one category of excessive work
can be properly excised or severed from the notice, whereas another category
attributable to an erroneous interpretation should not be excisable or

A severable and thus renders the entire notice bad. I of course accept that a notice which requires excessive work, whatever the reason may be for the excess, places the owner of a property in a dilemma. He can either, without further ado, comply with the unjustifiable requirement and thereby avoid any risk of an application compulsorily to acquire his property or he can take the following alternative steps. Under section 114(6) he may, within 28 days after the service of the compulsory purchase order, of which the authority seeks B confirmation from the Secretary of State, apply to a magistrates' court for an order staying further proceedings on the compulsory purchase order. If the court is satisfied that reasonable steps have been taken for properly preserving the building, the court shall make an order accordingly.

Additionally or alternatively, the owner, after service of the compulsory purchase order, for which confirmation is sought, may enter an objection to C the Secretary of State, who will then hold an inquiry. The compulsory purchase order may then be confirmed if the Secretary of State is satisfied that reasonable steps are not being taken for properly preserving the building. Even then, as stated above, he will not make or confirm a compulsory purchase order unless satisfied that it is expedient to make provision for the preservation of the building and to authorise its compulsory acquisition for that purpose (see section 114(4)). Mr Barnes, and Mr Sullivan QC for the D respondents, both accepted that before the magistrates or at the inquiry it was not essential for the owner to have actually carried out any work, so long as he could establish that he had taken reasonable steps with a view to carrying out such work. For example, he may have instructed surveyors to advise him as to which of the items in the repairs notice could fairly be said to be reasonably necessary for the proper preservation of the building, and having received their report and found that only certain items fell within that category, E instructed them to carry out such work after the proceedings before the magistrates or the inquiry had terminated.

Whichever is the reason for the owner's quandary, these options are open to him. Moreover where his contention is that neither the local authority nor the Minister is entitled in law to construe "works. . . reasonably necessary for the proper preservation of the building" in the manner in which they have done, he has the additional right of appeal from the Crown Court to the Divisional F Court by way of case stated, and from the decision of the Secretary of State to the High Court under section 23 of the Acquisition of Land Act 1981.

I therefore conclude that, so long as there is not inextricably mingled in the repairs notice works which have not the character of work of preservation, such works can properly be excised from the repairs notice, leaving the notice valid as respects those works which are reasonably necessary for proper G preservation of the building.

Since it is common ground that the applicant had not carried out even those 14 out of the 20 items in the repairs list which were undoubtedly reasonably necessary for the proper preservation of the building, the Secretary of State was entitled to confirm the compulsory purchase order. It is also common ground that he was entitled to conclude that it was expedient to make provision for the preservation of the building and to authorise its compulsory H purchase for that purpose. Accordingly, for the above reasons and those given by my noble and learned friend Lord Bridge of Harwich I would dismiss this appeal with costs.

LORD LOWRY: My Lords, I have had the advantage of reading in draft the speech of my noble and learned friend Lord Bridge of Harwich. I agree with it and, for the reasons which he gives, I, too, would dismiss the appeal.

Appeal dismissed with costs.

Camden London Borough Council *v* Secretary of State for the Environment and another

QUEEN'S BENCH DIVISION
Mr Malcolm Spence QC (sitting as a deputy judge of the Queen's Bench Division)
March 3 1989

Use Classes Order 1987 — Class B1 — Local plan policy protective of light industrial employment — Planning permission for light industry subject to conditions restricting use to light industrial — Application to use premises without restrictions refused — Appeal allowed on basis that restrictions frustrated government policy behind the Use Classes Order — Whether conditions valid — Whether national policy must take into account local circumstances — Whether retention of light industry is material consideration — Whether clear and intelligible reasons for decision A

The applicant local planning authority adopted a new local plan in May 1987. B The plan had policies intended to retain light industrial employment. In accordance with those policies, in granting two planning permissions to the second respondents, PSP (Nominees) Co Ltd, for the use of certain premises for light industrial purposes, they imposed conditions restricting the use of the premises to light industry and for no other purpose within class B1 of the Town and Country Planning (Use Classes) Order 1987 (SI 1987/764). The Secretary of State for the Environment allowed an appeal by the second respondents C against a refusal by the applicants to allow the use of the premises without the restrictions.

In allowing the appeal, the inspector based his decision on the guidance in Circular 13/87, concluding that the applicants' attempt to retain the distinction between light industry and offices would frustrate the aim of government policy following the introduction of the Use Classes Order.

Held The application was allowed and the decision of the Secretary of State D quashed.

The paragraph in the inspector's report which stated that the local planning authority's decision was contrary to government policy, as expressed in Circular 13/87, was the crucial determining paragraph. It dismissed the local planning authority's case concerning the protection of light industrial employment out of hand because it did not comply with government policy. The inspector had used a broad brush approach; this approach cannot be used E in every case. Not only did government policy expressly admit of exceptions, but also it would be manifest that whereas the broad aims were to foster enterprise, there may be parts of the country where enterprise would not be fostered. The decision letter had not taken into account the local planning authority's employment case and it gave no reasons for rejecting that case: see p 84.

Cases referred to in the judgment F

Centre 21 Ltd v *Secretary of State for the Environment* [1986] 2 EGLR 196; (1986) 280 EG 889; [1986] JPL 915, CA

Edwin H Bradley & Sons Ltd v *Secretary of State for the Environment* [1982] EGD 1280; (1982) 264 EG 926

ELS Wholesale (Wolverhampton) Ltd v *Secretary of State for the Environment* [1987] JPL 844, CA

Poyser and Mills' Arbitration, Re [1964] 2 QB 467; [1963] 2 WLR 1309; [1963] 1 All ER 612; *sub nom Poyser* v *Mills* [1963] EGD 421; (1963) 185 EG 609

A *Stringer* v *Minister of Housing and Local Government* [1970] 1 WLR 1281; [1971] 1 All ER 65; (1970) 68 LGR 788; 22 P&CR 255; [1971] JPL 114

Surrey Heath Borough Council v *Secretary of State for the Environment* (1986) 53 P&CR 428; [1987] JPL 199

Westminster City Council v *Great Portland Estates plc* [1985] AC 661; [1984] 3 WLR 1035; [1984] 3 All ER 744; (1984) 50 P&CR 34; [1985] JPL 108, HL

Appeal under section 245 of the Town and Country Planning Act 1971

B This was an appeal under section 245 of the Town and Country Planning Act 1971 against a decision of the Secretary of State for the Environment to allow an appeal against a refusal by the applicant local planning authority to grant planning permission.

Simon Pickles (instructed by the chief executive and town clerk, Camden London Borough Council) appeared for the applicant local planning authority.

C *David Holgate (instructed by the Treasury Solicitor) appeared for the first respondent, the Secretary of State for the Environment.*

Richard Langham (instructed by Frank Charlesly & Co) appeared for the second respondents, PSP (Nominees) Co Ltd.

The following judgment was delivered.

MR MALCOLM SPENCE QC: In May 1987 the London Borough of Camden adopted a new local plan after a long public inquiry. It had in it
D policies intended to retain light industrial employment. Only one month after the adoption of the local plan the new Town and Country Planning (Use Classes) Order 1987 (SI 1987/764) came into operation. It combines into one class, namely class B1, use for any of the following purposes: offices, research and development, and light industry. The effect of this is, in the light of section 22(2)(*f*) of the Town and Country Planning Act 1971, that planning
E permission is not required for a change of use between any of these uses. Accordingly the London Borough of Camden, in order to give effect to their local plan policies, have been in the habit of imposing a condition restricting the use of premises solely to light industry and for no other purpose within class B1.

However, in May 1987 the Department of the Environment issued Circular 13/87 relating to the new Use Classes Order and I shall read para 3. After
F referring to the aim of the new order as being two-fold, it goes on:

> It serves no-one's interest to require planning permission for types of development that generally do not damage amenity. Equally, the Secretaries of State are in no doubt that effective control must be retained over changes of use that would have a material impact, in land-use planning terms, on the local amenity or environment.

G Para 4 is also of a relevance, but I shall read just the last sentence:

> The aim of these new classes is to allow more flexible use of premises and thus to foster enterprise where that can be achieved without significant adverse effects on the environment for local amenity.

I must also refer to the Planning Policy Guidance Note no 4*, which was produced in January 1988, at para 16, where it is stated:

H

> Save in exceptional circumstances, conditions should not be imposed which restrict future changes of use which, by virtue of the Use Classes Order 1987, would not otherwise constitute development. The Secretaries of State would regard such conditions as unreasonable unless there was clear evidence that the uses excluded would have serious adverse effects on the environment or amenity, not susceptible to other control.

* Entitled *Industrial and Commercial Development and Small Firms.*

In the present case the London Borough of Camden had granted two A planning permissions authorising use of the subject premises [3-10 Warren Mews and 1-3 Fitzroy Mews, W1], comprising some 16,000 sq ft, for light industry. Both permissions were subject to conditions of the kind that I have mentioned. The second respondents applied for permission to use the premises for class B1 purposes without the restriction of the conditions. In effect, that was their application, though it was put somewhat differently upon the form. B

The London Borough of Camden refused planning permission for this reason:

> It is considered that the buildings are suited to providing accommodation for light industrial purposes in accordance with the Council's policies for the Fitzrovia area. Any widening of the scope of uses in these buildings would undermine the Council's policy to try and maintain a balance of uses in the C Fitzrovia Area and prevent any one use becoming dominant to the detriment of the mixed use character of the Fitzrovia area.

The second respondents appealed and the matter was dealt with by way of written representations. An inspector made a report and I need to read much of this. At para 5, under the heading "Appraisal" he wrote:

D

> A decision in this case essentially rests on the importance which is attached to the retention of light industrial employment in this part of Central London. The Council has clearly experienced some difficulty in dealing with this issue, because the Use Classes Order of 1987 was introduced one month after the adoption of the Borough Plan which contains policies to retain light industrial employment. The policies in the plan assume that the previous distinction between office and light industrial use would remain.

E

In para 6 he refers to the government's circulars, and then in para 7 he says:

> In the present case approval of the proposal would not adversely affect the local environment. Office use would be compatible with the character of the area, and create no loss of amenity for local residents or businesses. If amenity is the only test applied, the appeal must be allowed. However, the Council has suggested that a further consideration is appropriate in the particular circumstances of Central London. It believes that the higher rents F offered for offices will inevitably lead to the occupation of the appeal premises for office use and a consequent loss of light industrial floor space. The resulting change in the employment opportunities of local residents, and in the mixed use character of the area are said to constitute demonstrable harm and to provide a justification for maintaining a distinction between the two uses. The Council does not regard the appeal decision in the *Spelthorne* case referred to by the appellants as relevant, since in that case the borough G plan was out of date, whereas the Camden Borough Plan is only a year old.
>
> 8. I believe that the appeal premises are likely to be used, at least predominantly, for office purposes. Moreover, they are clearly capable of light industrial use since they are in the process of being modified for this purpose following a recent grant of planning permission. Therefore, if the loss of light industrial floorspace is considered to be of sufficient importance, the appeal should be dismissed.
>
> H
>
> 9. However, I have concluded that the Council's decision is contrary to Government policy, which has emphasised the need for local planning authorities not to interfere unduly with matters which go beyond strict land-use planning concerns. The aim of the introduction of the Business Use Class was to allow more flexible use of premises and to foster enterprise, where that can be achieved without significant adverse effects on the environment or local amenity. The Council's attempt to retain the distinction between light industry and offices in this case would frustrate the achievement of this aim.

A Having received that report, the Secretary of State, in his decision letter, referred to the council's reasons for refusal and the amplification in their statement. In para 4 he said:

> The Secretary of State agrees with the officer's appraisal of the case and accepts his recommendation. Therefore, for the reasons given by the officer, the Secretary of State allows the appeal . . .

B The London Borough of Camden challenged this decision under section 245 of the Town and Country Planning Act 1971. They say that the inspector and the Secretary of State, who adopted his reasons, have gone wrong in describing the matters raised as "being beyond strict land use planning concerns" and in doing so have failed to have regard to a material consideration. They also say that the Secretary of State and the inspector have failed to give clear and adequate reasons for the decision.

C It has to be said at the outset that this is an important case for the London Borough of Camden, whose policies as to offices and light industry would be rendered nugatory if in every case permission had to be granted in the light of the last two sentences of the inspector's conclusions. The heart of their case lies in para 4.12 of their representations. I read just one sentence:

> It is hoped that this Statement has also made it clear that the current state of
> D the commercial property market in Central London is such that the Council must secure new developments solely for light industrial use in order to protect and encourage a varied economic base best suited to offering wide employment opportunities.

There was much else in their representations to that effect. It was all based on their recently adopted objectives and policies. For example, one of the objectives was "to ensure a wide variety of employment opportunities in the
E area", and their policy PY37: "The mixed use character of the area will be conserved with the council actively protecting and promoting residential, light industrial and shopping uses as a priority . . .". Then of particular relevance among the employment policies is EM1: "(iii) to maintain a wide range of job types but to give priority to economic activities which" — and I just note some of them — "provide jobs for the unskilled and semi-skilled; . . . provide new jobs for unemployed residents in the borough . . .".

F The London Borough of Camden made it particularly clear in their representations that it was Fitzrovia that they were concerned about in the context of this planning appeal. At para 4.4 of their representations they said:

> The Council's opinion, however, is that in the case of an area such as Fitzrovia it can only aggravate the changes already referred to . . . and radically affect the character of the area.

G They then went on to describe how in areas such as Fitzrovia there would be the increase in rents of the kind to which I have already referred by reading para 7 of the inspector's report.

Although they did not expressly mention Planning Policy Guidance Note no 4, para 16, in their representations, it is plain that a case of this kind is capable of being made out in the light of the opening words of the paragraph,
H which I have read, "save in exceptional circumstances".

Mr Pickles, who appears on behalf of the London Borough in this court, makes four submissions that are not disputed: first, that the original conditions were lawful. Second, that the application of national policies in any particular area must take into account local circumstances. He cites for that proposition *Surrey Heath Borough Council* v *Secretary of State for the Environment* (1986) 53 P&CR 428 and also *ELS Wholesale (Wolverhampton) Ltd and Crownbrae Ltd* v *Secretary of State for the Environment and*

Cheltenham Borough Council [1987] JPL 844. In particular he drew my A
attention to the judgment of May LJ, where on p 846 he said:

> Before turning to the circulars themselves, Kennedy J in the recent case of *Surrey Heath Borough Council* . . . pointed out what had been pointed out before, but which needed stressing: namely, that the circulars were issued nationally and, therefore, had to be applied in widely differing circumstances throughout the country. For that reason, let alone for any others, the wording of a circular was not to be, and could not be, applied in a particular B
> case as if it were a statutory instrument or a section of statute.

It is particularly pertinent, in my judgment, to bear those words in mind when considering the circumstances of the present case.

Mr Pickles' third submission was that the retention of light industrial uses to protect the balance of uses in a particular area directly concerns the use of land, serves a planning purpose and is therefore a land use planning C
consideration within section 29(1) of the 1971 Act. For that proposition he cited the well-known case of *Stringer* v *Minister of Housing and Local Government* [1970] 1 WLR 1281 and also the House of Lords case *Westminster City Council* v *Great Portland Estates plc* [1985] AC 661.

He drew particular attention to the speech of Lord Scarman. I shall refer to it only briefly, but it is of considerable interest in the sense that in Westminster there are policies that are remarkably similar to the policies, to which I have D
already referred, in Camden. Lord Scarman refers to these on pp 666 and 668. It is pertinent to observe that in the passage which is cited on p 668, giving the reasons for the policy, the main point made is that these uses are vulnerable to pressure from other more financially profitable uses. The relevance of the reference to these policies in the context of that case is to underline the submission that they are a relevant land use planning consideration.

Mr Pickles' fourth submission was that this retention of light industrial uses E
was a planning consideration which is material to this appeal. It was plainly judged to be material by the inspector because he expressly referred to it in paras 5 and 8 of his report.

Mr Pickles made a fifth submission, and this was a submission that was in dispute, namely, that this planning consideration, to which I have just referred relating to the retention of light industrial uses, was wrongly not taken into account. Then, as I have said, by virtue of the amendment to the F
notice of motion he submitted that the inspector and Secretary of State had therefore failed to give clear and intelligible reasons.

Mr Holgate, who appears on behalf of the Secretary of State, and Mr Langham, who appears for the second respondent and who has supported Mr Holgate's submissions, stress that the light industrial floorspace argument was taken into account as one can plainly see from para 8 of the report and that G
the first sentence of para 9 does not purport to decide that these factors are irrelevant. Mr Holgate says that this first sentence is a reference to the last sentence of para 3 of Circular 13/87 and in that regard he seems to me to be correct, particularly by reference to the use of the words "land use planning terms".

It is of some importance that I should record that Mr Holgate also says, on behalf of the Secretary of State, that para 3 of the Circular 13/87 is not H
intended to restrict the ability to refuse applications of this kind merely by reference to amenity and the environment. A local planning authority are still entitled to refuse such planning applications on the basis of, for example, employment considerations, or indeed the very kind of reasons which the London Borough of Camden advanced in this case. Moreover, Mr Holgate says that, in the first sentence, the reference to "land-use planning concerns" is not a reference to the London Borough of Camden's employment case.

In my judgment, this crucial determining paragraph has not dealt with the London Borough of Camden's employment case, nor indeed para 8 of the conclusions themselves where the point has in effect been put up for consideration. The paragraph dismisses Camden's case concerning employment out of hand because it does not comply with government policy. As expressly stated in the second sentence of para 9, the new class B1 was intended to foster enterprise, and, as stated in the last sentence of para 9, the council's policy would frustrate the achievement of this aim. This is a broad brush approach which could be applied in every case. That cannot be right. Not only does the government policy expressly admit of exceptions, but it is also manifest that, whereas the broad aim is to foster enterprise, there may be parts of the country where enterprise would not be fostered by this government policy or indeed may actually be retarded. Each case, including a case of this kind, always has to be considered on its individual merits. Nowhere do I see anything in this reasoning to the effect that Camden's assertion that there are special factors appertaining to Fitzrovia are in truth no different from circumstances obtaining elsewhere; nor do I see it anywhere said that there is in truth no case, as Camden were pleading, for treating the circumstances in Fitzrovia as an exception to the general policy.

Mr Holgate said that the conclusions would have been unimpeachable if words had been added to the effect that the case put forward by the London Borough of Camden was of insufficient weight. He said that I can infer that that is what was in the mind of the inspector from all the surrounding circumstances. In my judgment, it would not be adequate to answer the case being made by the London Borough of Camden about the special circumstances in Fitzrovia, particularly having regard to the need to provide employment for the unemployed and the unskilled, merely to say that their case put forward was of insufficient weight.

I uphold this application for two reasons: first, although the decision letter has clearly mentioned Camden's employment case in para 8, it has not taken it into account when making the actual decision but rejected it by reference to general policy. Second, the letter gives no reasons, not merely no intelligible reasons, for rejecting the employment case.

Mr Holgate has helpfully referred me to the recent Court of Appeal decision in *Centre 21 Ltd* v *Secretary of State for the Environment* [1986] 2 EGLR 196, where on p 199 Glidewell LJ rehearsed the cases relating to the giving of reasons, including *Westminster City Council* v *Great Portland Estates plc*, where there is reference to the decision of Megaw J in *Re Poyser and Mills' Arbitration* [1964] 2 QB 467, where in particular he said "that there must be something 'substantially wrong or inadequate' in the reasons that are given". That is even though, as Glidewell J (as he then was) said in *Edwin H Bradley & Sons Ltd* v *Secretary of State for the Environment* (1982) 264 EG 926, they may be briefly stated.

In my judgment, the reasons given here were wholly inadequate. For these reasons the decision of the Secretary of State will be quashed.

Application allowed with costs. Decision of the Secretary of State quashed.

Wyre Forest District Council *v* Secretary of State for the Environment and another

COURT OF APPEAL
Slade and Lloyd LJJ and Sir George Waller
March 21 1989

Application for planning permission — Proposals contrary to development plan — Appeal — One inspector hearing appeal — Second inspector holding local plan inquiry — Inspector granted planning permission on appeal — Local plan inquiry inspector recommended site remain as playing-fields — Whether planning appeal inspector entitled to reach his decision A

In April 1986 the Order of the Holy Trinity Convent ("the Order") applied to Wyre Forest District Council, the local planning authority, for outline planning permission for the residential development of certain school playing-fields. The council refused consent, and one of the grounds of refusal was that the site was zoned on the local plan for the purposes of a school and grounds. The Order appealed against that refusal and in February 1988 the Secretary of B State for the Environment, by his inspector, allowed the appeal and granted planning permission on three main grounds: the unlikelihood of the site reverting to school playing-field use; the local need for housing land; and the likely environmental impact.

At about the same time another inspector was holding an inquiry into the local plan. In March 1988, this inspector, in his report to the council, recommended that the site should remain allocated on the local plan as C "education establishments and grounds". The council's application under section 245 of the Town and Country Planning Act 1971, to have the decision to grant planning permission quashed, was allowed by His Honour Judge Marder QC (sitting as a judge of the High Court) (September 29 1988) primarily on the basis that the planning appeal inspector's decision not to defer his decision until after the local plan inspector's report was known, or not to dismiss the appeal on its merits, was *Wednesbury* unreasonable. The Order and Secretary of State for the Environment appealed that decision. D

Held The appeal was allowed.

Had the planning appeal inspector decided that it would have been premature to have granted planning permission before the local plan report, it was doubtful that any court could properly have interfered with such a decision: see p 93A. However, the inspector did not so decide and he was entitled to reach the decision he did for the following reasons: (1) the E availability of housing land was crucial to his decision; he had carefully considered the evidence and concluded that there was no identified five-year supply in the area, a decision reached on sufficient material and one that could not be described as perverse; (2) once he had identified a shortfall in the five-year supply of housing land, there was the presumption in favour of granting planning permission in Circular 15/84; (3) the uncertain prospects of the site reverting to school use; (4) the general need to avoid delay in planning matters as recommended in PPG1; and (5) the uncertainty as to the timing and F outcome of the local plan inspector's report: see pp 93B-94C and 95.

Cases referred to in the judgments

Allen v *Corporation of the City of London* [1981] JPL 685
Arlington Securities Ltd v *Secretary of State for the Environment* [1989] 1 PLR 14
Associated Provincial Picture Houses Ltd v *Wednesbury Corporation* [1948] 1 KB 223; [1947] 2 All ER 680; (1947) 45 LGR 635, CA
Davies v *Hammersmith and Fulham London Borough Council* [1981] JPL 682, CA

Appeal against a decision of His Honour Judge Marder QC

This was an appeal against a decision of His Honour Judge Marder QC (sitting as a judge of the High Court) (September 29 1988), who had allowed an application by the respondents, the Wyre Forest District Council, under section 245 of the Town and Country Planning Act 1971, to challenge a decision of the appellant Secretary of State for the Environment, who had, by his inspector, allowed an appeal under section 36 of the 1971 Act against a refusal of planning permission by the respondents.

Christopher Lockhart-Mummery QC and Christopher Katkowski (instructed by Arnold Fooks Chadwick & Co) appeared for the first appellants, the Order of the Holy Trinity Convent.

David Pannick (instructed by the Treasury Solicitor) appeared for the second appellant, the Secretary of State for the Environment.

David Spens (instructed by the solicitor to the Wyre Forest District Council) appeared for the respondents.

The following judgments were delivered.

SLADE LJ: There are before the court appeals by the Order of the Holy Trinity Convent and by the Secretary of State for the Environment from an order of His Honour Judge Marder QC, sitting as a judge of the High Court, made on September 29 1988 in a planning matter. The first appellants (whom I will call "the Order") are the trustees of a charitable trust concerned with education in the neighbourhood of Kidderminster. Until recently they ran a school known as the Holy Trinity Convent School at Kidderminster. However, they then found it necessary to close it. A group of parents successfully negotiated to acquire and maintain the school as a charitable trust. It was therefore arranged that the group should acquire part of the land owned by the Order comprising the school buildings and a tarmacadam games surface. However, the group was not able to afford to buy the school playing-fields. The Order regarded it as their duty as trustees, so far as consistent with satisfying the requirements of the school, to maximise the value of their land not required upon the disposal for the purposes of the new school.

By an application dated April 28 1986 and received on May 12 1986, the Order accordingly applied to the Wyre Forest District Council ("the planning authority"), the respondents to these appeals, for outline planning permission for the residential development of the playing-fields comprising some 1.62 hectares of land ("the site").

On November 18 1986 the planning authority gave notice of refusal of the application on the following grounds:

> 1. The site is zoned on the Approved Kidderminster Review Town Map for the purposes of a School and Grounds. The proposal would be contrary to this zoning and the principles underlying it and would prejudice the use of this area for these purposes.
>
> 2. The proposed new access and requisite visibility splays would seriously harm the character and appearance of this particularly attractive area.
>
> 3. The development proposed if carried out would severely restrict the ability of the present owners of the school to provide adequate playground and playing-field space.

On May 30 1987 the Order gave notice of appeal to the Secretary of State for the Environment from the planning authority's refusal of their application.

The Secretary of State then appointed an inspector, Mr B E Partridge, to determine the appeal. The planning authority prepared a written statement of submissions which were to be made to the inspector at the ensuing inquiry, and these included the following:

> 6.1 It will be submitted that the development proposed by the Appellant is clearly contrary to the zoning of the Appeal site on the approved Kidderminster Review Town Map, i.e. for the purposes of a school and grounds. A
>
> 6.2 It will further be submitted that the proposed development is equally contrary to the proposed zoning of the Appeal site for "educational establishments and grounds" in the Wyre Forest Urban Areas Local Plan. A Public Local Inquiry into the objections to the Local Plan is currently being B
> held and, in relation to the Appeal site, the Appellant has objected to the proposed zoning in the Local Plan and has contended that the Appeal site should be allocated as a new housing site. As the Local Plan Inquiry Inspector's findings about objections to the Local Plan may not be available to the Council until perhaps the early spring of 1988, it will be argued by the Council as part of their response to the Grounds of Appeal (paragraph 7.1 below) that the Appeal proposal is premature. The Council will submit evidence that there is no overwhelming need for additional housing land C
> which would justify development of this attractive and valued open space. The Council will submit that the Appeal site should remain designated and be used for the purpose specified in the Town Map and the Local Plan.

These references to the local plan perhaps require some brief explanation. Broadly, a local plan consists of a written statement formulating the proposals of a local planning authority for the development and other use of land in the D
part of their area covered by the plan. The plan, when prepared, is put on deposit. After the expiry of the period afforded by statute for making objections to the plan, a local plan inquiry will be held into the objections by one of the inspectors appointed to hold such inquiries. The inspector so appointed, having concluded such inquiry, will report to the planning authority and will make what he considers appropriate recommendations. In so doing (unlike an inspector hearing an appeal under section 36 of the Town E
and Country Planning Act 1971 from the refusal of planning permission by a local planning authority), he will not be performing a function of the Secretary of State. He will be affording assistance to the local planning authority, which will have the right and duty to decide whether or not to give effect to his recommendations, and if so with what modifications (if any). The local plan inspector, in making his report, will himself decide nothing.

Not infrequently, however, an inspector hearing an appeal from the refusal F
of planning permission by a local planning authority ("a section 36 appeal") may be told either that a local plan inquiry relating to an area including the appeal site is shortly to take place or that a report following the holding of such an inquiry is in the course of preparation.

An inspector determining a section 36 appeal is bound to have regard to "the provisions of the development plan, so far as material to the application, G
and to any other material considerations": see sections 29(1) and 36 (5) of the 1971 Act. The provisions of an emerging local plan, or the fact that the report of a local plan inspector is shortly to be made, must clearly be capable of being a material consideration to be taken into account in determining the appeal. Considerations of this nature may make it necessary for the inspector hearing the section 36 appeal to determine whether it is appropriate to deal with the application at that stage or whether he should treat the application as H
premature. Nevertheless, it is plain that the existence of an emerging local plan or of a pending report of a local plan inspector does not necessarily and as a matter of principle prevent him from proceeding to hear the appeal on its merits any more than it will necessarily make it wrong as a matter of principle for a planning authority to perform their functions by finally dealing with a planning application in such circumstances: see and compare *Allen* v *Corporation of the City of London* [1981] JPL 685 and *Davies* v

A *Hammersmith and Fulham London Borough Council* [1981] JPL 682. In such a case the inspector hearing the section 36 appeal will have a wide discretion in deciding how best to proceed.

The existence of such discretion is confirmed by Circular 22/84 dated September 7 1984 and issued jointly by the Department of the Environment and the Welsh Office. Para 1.12 explicitly states:

> B . . . the fact that a local plan is in the offing or that there are proposals for alteration or repeal or replacement of a structure plan or a local plan is not, in itself, a reason for refusing planning permission.

More recently, Planning Policy Guidance Notes PPG1* — dated January 1988, and issued jointly by the same two bodies, has emphasised the need to avoid "delay and uncertainty" and the need to achieve "promptness, relevance and efficiency" in planning matters. Para 6 states as a general
C principle that "planning applications should normally be decided within eight weeks" and that for inquiry cases decided by inspectors the aim is to achieve a "median time" of 17 weeks from the lodging of the appeal to the issue of the decision.

Before turning to the decision of the inspector on the section 36 appeal in the present case, it will be convenient to refer to another circular, no 15/84, dated
D July 4 1984, issued by the Department of the Environment, headed "Land for Housing", which in the first sentence expresses the concern of the Secretary of State that "the planning system should provide an adequate and continuing supply of land for housing". Paras 2 and 3 of Annex A to this circular, so far as material, provide as follows:

> *Supply of Land*
>
> 2. For any given area, the availability of land for housing will be governed
> E primarily by the policies set out in the development plan. However development plan policies do not in themselves ensure that the housebuilding industry can produce the houses needed. For that there must be an adequate and continuous supply of land, with planning permission, suitable and available for immediate development, and situated where potential house buyers are prepared to live. This is why DOE Circular 9/80 (WO Circular 30/80) asks planning authorities to identify specific sites providing a five year
> F supply of housing land in accordance with structure plan policies and, where the authority is approached by the housebuilding industry, to discuss with the industry whether that land is genuinely available for development.
>
> *Policy in the Absence of an Identified Supply of Land*
>
> 3. In the absence of such an identified five year supply there should be a presumption in favour of granting permission for housing except where there are clear planning objections which in the circumstances of the case outweigh
> G the need to make the land available for housing. The relevant factors should be apparent from the development plan . . . .

In para 13 of this circular, the Secretary of State emphasised that a five-year supply of housing land is a minimum requirement.

In the present case the inspector appointed by the Secretary of State held an inquiry into the Order's section 36 appeal on November 30 and December 1 1987. Meantime, on various dates in October and November, the quite
H separate local plan inquiry had been proceeding before another inspector; it continued in December.

Submissions, on the lines foreshadowed by the planning authority's written submissions (to the effect that the Order's application for planning permission was premature) were made to the Secretary of State's inspector. However, by a full and careful decision letter dated February 12 1988 he

* Entitled *General Policy and Principles.*

rejected them. He allowed the Order's appeal and granted planning permission for the residential development of the site, subject to certain conditions. In para 2 of his decision letter, the inspector identified what were in his view the three main issues:

> . . . The first is the need for the appeal land to revert to school playing-field use, and the likelihood of its so doing in the event of either the refusal or the grant of planning permission for residential purposes. The second is the need for the land for housing purposes, in the context of the requirement for a 5-year housing land supply and of current local plan proposals for further land allocations for that purpose. The third is the likely environmental impact of the proposed development, in visual and in traffic terms.

In paras 3 and 4 the inspector referred to points drawn to his attention respectively by the Order and the planning authority in the context of the playing-fields issue. While the purchasers of the school had previously fully acquiesced in the retention by the Order of the appeal site and in their seeking permission to develop it for residential purposes, they had now opened an appeal for funds so that they could buy it and use it for its former purpose. "However", stated the inspector, "they [the Order] were bound to sell the land at full market value, and therefore still sought a residential permission. If this was forthcoming, they may be able to sell to the school trustees, or to some other purchaser, subject to the Charity Commissioners' consent. If permission were refused, the Order would retain the land. Dismissal of the appeal would not therefore mean that there was any reasonable prospect of the land being made available again for playing-field purposes". On the other hand, while the school had recreational facilities that amply complied with local authority school standards, the council considered that there was still a need for permanent playing-fields for the school. The inspector said:

> . . . The appeal land was allocated on the current statutory plan, the Kidderminster Review Town Map, for educational purposes. When the Wyre Forest Urban Areas Local Plan was being prepared it appeared that the land had been lost to the school on its change of ownership, and residential use was suggested. It now seemed that there was a good prospect that the school, which had launched an appeal for funds and had raised a substantial proportion of the necessary purchase price, would be able to acquire the land, which was ideally suited to resume its previous use. In the light also of an indication that wider community use would be permitted the council now felt that playing-field use was preferable to the previously proposed residential use, and had accordingly changed the allocation on the Local Plan to "education establishments and grounds". An objection to this allocation had been heard in the course of the recent Local Plan Inquiry.

In the light of these submissions, the inspector (in para 5) accepted that on the playing-fields issue "taken for the moment in isolation" the case had been made out for the renewed use of the land as playing-fields for the school. He said:

> . . . Although in theoretical terms the school may have enough recreational space already, there is clearly advantage in having suitable permanent grassed areas immediately adjacent to the buildings, and the school trustees are obviously seriously intent on acquiring these extra facilities. Whether there is a reasonable prospect of land being made available for this purpose seems to me inconclusive from the evidence; from the conflicting submissions on this matter I take the view that this prospect will depend on further discussions between the parties, which are likely to continue whatever the result of the present appeal. Accordingly I conclude that playing-field use would be an appropriate one in planning terms, and that the strength of the case for that use must go into the balance against the competing residential use which has been applied for.

A The inspector then dealt with "the housing land availability issue" (which
was of importance because it would determine whether the usual presumption
set out in Circular 15/84 would apply). As to this issue, he expressed his
findings and conclusions thus:

> 6. Despite differing conclusions, there was some consensus between the
> principal parties on this issue. From the figures submitted by the council it
> B was clear that at 1 April 1987, and using the strict residual method of assessing
> availability, there was only 3.8 years' supply of housing land. The Council
> argued, however, first that by adding the sites allocated in the deposited
> Local Plan, this could be increased to 5.1 years' supply, and secondly that by
> using the "past completions" method referred to in Departmental News
> Release 220, which they said was appropriate to Wyre Forest, an ample
> supply of 11.4 years was found to be available. For the appellants however it
> was said that it was not appropriate to take account of the Local Plan
> C allocations, many if not most of which would not be available within 5 years,
> and were much less suitably located for housing development than the appeal
> site, and that the "past completions" method was not the proper assessment
> method.
>
> 7. In my opinion there is no convincing reason why the residual method of
> calculating the housing land supply, which News Release 220 indicated
> should be the usual method to apply, should be abandoned in this case.
> D Although the housebuilding rates in Wyre Forest have been significantly
> lower in the past than in other districts in Hereford and Worcester, there is no
> real indication of the reason for this, and the recent trend has been upwards.
> The council have not demonstrated to my satisfaction, accordingly, that low
> demand is likely to remain a factor important enough to justify departing
> from the usual methods of assessing housing land supply.
>
> 8. Although it was argued on the appellants' behalf that no account should
> E be taken of the Local Plan allocations, since these were still the subject of
> many objections and might well not appear in the finally approved and
> adopted plan, I have in mind that the plan has reached a late stage in its
> development, and its provisions must therefore carry substantial weight in
> my consideration of this case. By the council's own figures, even if all the
> allocations were to be approved there would only be 5.1 years' supply, or 5.4
> years if the figures were to be modified by the County Council's amended
> annual housing needs for 1991-1993. The position at its best is therefore very
> F close to the margin, and bearing in mind that the 5-year supply suggested by
> Circular 15/84 is expressed to be a minimum only, and that there is an
> obvious possibility that not all the Local Plan sites will become available
> during the next 5 years, on the evidence produced by the appellants, I am
> satisfied that at the present time there is no identified 5-year housing land
> supply in Wyre Forest, and that the presumption in favour of granting
> permission set out in Annex A to the Circular should operate.

G Having referred to certain matters relating to the environmental aspect, in
para 12 of his decision letter the inspector expressed his conclusion thus:

> My conclusion is that although school playing-field use is one that would be
> entirely appropriate, and would avoid having to build over an attractive open
> area, there is no convincing real need for that facility sufficient to override the
> identified housing need in the district which the development under appeal
> H would help to meet. I have taken into account the suggestion that a
> permission now might prejudice or conflict with any recommendations
> arising from the recent Local Plan Inquiry, but am of the view that a decision
> in this case can be determined on its own merits and need not await decisions
> in the wider context of the local plan.

The inspector, by his decision letter of February 12 1988, accordingly
allowed the Order's appeal from the planning authority's decision of
November 18 1986.

In the event, the local plan inspector signed a written report on March 18 A 1988 stating that he agreed with the council's decision and recommended no change in the use allocated for the site in the local plan, "education establishments and grounds".

The decision of the Secretary of State's inspector given by the letter of February 12 1988 fell to be treated as a decision of the Secretary of State under section 36 of and Schedule 9 to the Town and Country Planning Act 1971 ("the 1971 Act"). On March 23 1988 the planning authority issued a notice of B motion joining as respondents the Secretary of State and the Order and seeking an order that the decision of February 12 1988 allowing the Order's appeal against the planning authority's decision of November 18 1986 refusing outline planning permission for the residential development of the site should be quashed.

On September 29 1988, His Honour Judge Marder made an order in these C terms. Both the Order and the Secretary of State now appeal from the judge's order.

The judge, in giving judgment, rightly reminded himself that the planning authority could challenge the inspector's decision to grant outline planning permission only on the narrow grounds provided for by section 245 of the 1971 Act and that the burden on the challenger was a heavy one based, as it was, on unreasonableness in the *Wednesbury* sense. D

Of the three grounds of appeal argued before the learned judge, the first two represented objections to the manner in which the inspector had dealt with the important question of the adequacy or otherwise of the supply of housing land in Kidderminster. The judge, in my opinion rightly, rejected these objections, holding that they concerned matters of judgment falling within the inspector's "competence and expertise". There is no respondent's notice seeking to resurrect these first two grounds. E

The third ground of appeal argued before the learned judge, upon which he ultimately decided that the appeal succeeded, was that (p 15 of the transcript):

> . . . it was quite unreasonable for this inspector, in the circumstances of this case, not to do what had been suggested to him, i.e. either to refuse planning permission on the ground that it was premature to release this land for housing development in advance of the inspector's conclusions in the context of the Local Plan inquiry, or to defer a decision to await the light that his F colleague's report was bound to throw both on the general question of housing land availability in the district and on the necessity for the development of this site as opposed to other possibilities in order to meet any shortfall there may have been.

The inspector was not apparently invited to adjourn the inquiry at the hearing. In the light of the opposing submissions made to him, there were for G practical purposes four courses open to him, namely: (a) to dismiss the Order's application for planning permission, on a "without prejudice" basis, on the grounds that it was premature; (b) to defer his decision until after the local plan inquiry; (c) to dismiss the Order's application on its merits; or (d) to proceed to grant the planning permission sought. He considered that course (d) was the appropriate one and followed it. On the face of it, this was as much a matter of judgment for his competence and expertise as was the question of H the adequacy or otherwise of the supply of housing land in Kidderminster. However, Mr Spens, on behalf of the planning authority, who no doubt presented his argument in the court below as attractively as he has presented it in this court, succeeded in persuading the learned judge that the inspector's decision to adopt course (d) as opposed to course (a) or (b) was in all the circumstances so unreasonable that no reasonable inspector could have made it. Before this court he has made the like submission, at the same time specifically disclaiming any submission that the inspector failed to take into account any material factor which he should have taken into account.

It is not submitted on behalf of the planning authority that there is a general rule that an inspector hearing a section 36 appeal must in all cases defer a decision or refuse planning permission on the grounds of prematurity merely because a local plan is in the course of preparation. As the judge himself said (at p 14D), any such proposition would be quite obviously too wide. However, Mr Spens submitted that when it came before the inspector the present case was a special one for three reasons which were accepted by the judge.

First, it was said, the question of housing land availability was a key issue. Though the inspector was eventually satisfied that at the present time there was no identified five-year housing land supply in Wyre Forest, he recognised that the situation "at its best" was "very close to the margin".

Second, it was submitted on behalf of the planning authority, on the evidence before the inspector it by no means followed that the site would be used for housing, even if planning permission were granted. The Order had a duty to obtain the best price they could for the site. The owners of the school wanted the land for playing-fields if they could afford it. Thus, to quote the judge's words (at p 7):

> . . . the decision whether or not to build houses on this site, and thereby contribute to any shortfall in housing land supply, would rest ultimately not on the district's needs for housing land but on the ability of the school's owners to buy the land for playing fields at the enhanced price that the benefit of planning permission would confer.

Third, it was pointed out, at the time when the inspector issued his decision letter, the report and recommendations of the local plan inspector were imminent. In the words of the judge (at p 8):

> . . . That inquiry and the report would, of necessity, be district wide and would deal in detail with whether or not the existing supply of housing land was adequate; whether or not further allocations were required to be made, and would presumably consider and advise the local planning authority if further allocations were required — where and on what scale they might best be made; and where objections had been taken to proposed allocations (including the allocation of this land for educational purposes) whether those objections were sustainable. All those are matters that were within the remit of the inspector for the purposes of the local plan.

As regards the second of these three points, there can never be any certainty that planning permission will be acted upon if granted; no one can be compelled to act upon it. As at February 12 1988, however, it could have been forecast with a fair degree of confidence that the planning permission, if granted, would be likely to be acted upon *unless* the trustees of the new school were able to raise the substantial funds which would be required to purchase the site; and no one knew whether they would be able to do so.

The first and third of Mr Spens' three points, which are closely connected with one another, are in my judgment by far the more important. In considering his decision, the inspector could and no doubt did have in mind that the report and recommendations of the local plan inspector, when produced, might well contain features of the nature referred to by the judge in the passage cited above, which could be of assistance to the inspector if he were to defer his decision, or could be material if the present application of the Order for planning permission were dismissed as being premature and a further application had to be made. If, in the exercise of his judgment, he had decided that in all the circumstances it would be premature to grant planning permission before the report and recommendations of the local plan inspector A were available, it might well have been a decision with which no court could properly have interfered. We were referred to an unreported decision, *Arlington Securities Ltd* v *Secretary of State for the Environment* (October 3 1988)* on the particular facts of which this court held that it had been open to the Secretary of State to treat an application for planning permission as premature, because its implications required investigation and consideration by the local plan process, and refused to interfere with his decision. B

However, without much hesitation, I respectfully differ from the learned judge's view that in the circumstances of the present case any reasonable inspector acting reasonably would have been bound to adopt course (a) or course (b) above. Notwithstanding the force of the factors relied on by Mr Spens, a number of factors were drawn to our attention either by Mr Lockhart-Mummery, on behalf of the Order, or by Mr Pannick, on behalf of C the Secretary of State, which could be said to point in the other direction, ie in favour of course (d) (which the inspector adopted). Of these, the more significant, in my judgment, were the following:

(1) As the inspector himself appreciated, the housing land availability issue was crucial to his decision. Having carefully considered the evidence before him relating to that issue, he made a finding that there was no identified five-year housing land supply in Wyre Forest. The planning authority had had the D full opportunity to place before him any evidence relating to that issue on a "district wide" basis which they thought relevant. He was satisfied that he had sufficient material before him to reach a conclusion on that issue and it is not suggested that that conclusion was perverse.

(2) Having reached that conclusion, he was, in my judgment, entitled also to regard as applicable the presumption in favour of the grant of planning permission, stated in Annex A to Circular 15/84. The learned judge said that it E was clear that "if it were not for what the inspector perceived to be a pressing shortage of housing land he would have upheld the local planning authority's refusal of this permission and dismissed the appeal". A little later he said (at p 13E-F):

> there was no reasonable basis upon which he could conclude that the need for housing land was such as to justify overriding the strong objections which he F had identified to loss of this open playing-field.

With respect, I do not agree. The learned judge appears to have overlooked the existence of the presumption, to which he never explicitly referred. The inspector was, in my view, fully entitled to have regard to this presumption once he had concluded that there was no identified five-year land supply — though the weight which he attached to it would have been a matter for him.

(3) The inspector was entitled to regard as uncertain the prospects of the site G becoming available for renewed use as playing-fields. (In the event, the solicitors acting for the trustees of the new school informed the Order's solicitors by a letter of August 16 1988 that it would not be possible for the trustees ever to buy the site.)

(4) The inspector would also have been entitled to have regard to the general need to avoid delay in planning matters recently and forcefully stated in PPG1, and the maximum period of 17 weeks therein recommended as a target. H The Order's appeal had in fact been lodged on May 13 1987, substantially more than 17 weeks before the issue of the inspector's decision letter.

(5) The inspector, in announcing his decision, (a) would not have known how many weeks would have to elapse before the local plan inspector issued

* Reported at [1989] 1 PLR 14.

his report; (b) would not have known how much new light the report when issued would shed on the issues which he, the inspector, had to decide; (c) would have appreciated that the report when produced would in any event bind no one and that it might be many months before it was known whether or not the planning authority were proposing to adopt its recommendations, and if so with what modifications.

I see the force of the submissions made on behalf of the planning authority and have some sympathy with the learned judge's approach to this case. However, particularly in the light of the five factors which I have listed, taken together, I think that the circumstances did not justify his substituting his own opinion for that of the experienced inspector who, having conducted a full inquiry and heard all the evidence, had concluded that the right course was to give an immediate decision in favour of the applicant Order. The inspector specifically directed his mind to the relevance of the local plan process in the context of his decision. In my view, he was fully entitled as a matter of discretion to determine the appeal when he did, in favour of the Order, instead of obliging them to wait for an indeterminate period until their application, or a renewed application, could be dealt with. Whether or not the court below or this court would itself have followed the same course, it was not a course which no reasonable inspector could reasonably have taken in all the circumstances.

I would accordingly allow this appeal. I would set aside the order of the learned judge, and restore the decision of the Secretary of State's inspector dated February 12 1988.

LLOYD LJ: Like Slade LJ I have sympathy with the judge's approach in this case. I also have sympathy with his conclusion, more especially as the local plan inspector, in his report dated March 18 1988, recommended that there be no change in the existing use. I quote from his report:

> For the Council it was argued that on educational grounds the use of the land as school playing-fields was preferable to residential use. There was no doubt that the present Trustees regarded the use of this land as essential for the operation of the school since they had opened an appeal to purchase the site and restore the land to its former use as playing-fields. The Council considered that adequate provision for future housing needs has been made elsewhere and that this site is no longer required for that purpose.
> I agree with the Council's decision and recommend no change in the use allocated for this site in the Plan.

As the judge pointed out, it does no credit to the planning system when different inspectors reach conflicting results within a few weeks or months of each other. But where judgment is concerned, there is always room for difference of opinion. The fact that the decisions of the two inspectors were conflicting does not mean that one or other was unreasonable in the *Wednesbury* sense, so as to justify intervention by the courts.

Having set out verbatim the crucial paragraph of the decision letter dated February 12 1988, the judge said:

> That conclusion is . . . so unreasonable in the circumstances of this case that I do not consider that a reasonable Inspector, having in mind the relevant facts and background that I have referred to, could have reached it.

There are two strands to be discerned in the judge's reasons. At p 13 he says:

> As I see it, there was no reasonable basis upon which he could conclude that the need for housing land was such as to justify overriding the strong objections which he had identified to loss of this open playing-field.

I find myself unable to agree that there was no reasonable basis for the inspector's conclusion. The balancing of the need for housing land against the sacrifice of an open playing-field was essentially a matter for the inspector's planning judgment. One may regret his decision. Other inspectors might have reached a different conclusion. But I do not see how it could be said that the decision was one which no reasonable inspector could reach.

The second strand in the judge's reasoning is to be found at p 14:

> . . . I consider that it was quite unreasonable for this inspector, in the circumstances of this case, not to do what had been suggested to him, ie either to refuse planning permission on the ground that it was premature to release this land for housing development in advance of the inspector's conclusions in the context of the Local Plan inquiry, or to defer a decision to await the light that his colleague's report was bound to throw both on the general question of housing land availability in the district and on the necessity for the development of this site as opposed to other possibilities in order to meet any shortfall there may have been.

I find this second way of justifying the judge's conclusion more convincing than the first. For the question whether to defer a decision, pending the publication of the local plan inquiry report, or for whatever other reason, does not involve the exercise of a planning judgment. The question is the same as it would be in any other field, and depends, not on any special skill or experience in the planning field, but on common sense.

Mr Lockhart-Mummery accepted that if the report of the local plan inspector had been, as he put it, "just round the corner", then it would have been sensible for the appeal inspector to wait; "it would", he said, "have been silly to do otherwise". This must be so, even though the recommendations in the local plan report would not have been binding on anyone. Since the question whether the local plan was "just round the corner" is one on which the court is equipped to form a view, I can imagine circumstances in which the court would be ready to intervene where an inspector refuses to defer his decision. The court would not be usurping the inspector's planning judgment.

But the question whether the inspector should have deferred his decision, though not involving any exercise of his planning judgment, was still a question for his discretion. When we are asked to disturb his discretion the same rules apply in this field as in any other. There is no rule that an inspector *must* defer his decision when a local plan report is awaited.

Are there, then, grounds on which we would be justified in interfering with the inspector's discretion in this case? The judge answered that question in the affirmative. In his view, the local plan report was *so* imminent, and *so* relevant to the question which the inspector had to decide, that any reasonable inspector would have waited. I sympathise, as I say, with that conclusion. But on the facts I take a different view. In my judgment, there were grounds here on which the inspector could exercise his discretion in the way he did, ie against deferring his report. Those grounds are fully set out in the judgment of Slade LJ. It would serve no purpose for me to repeat them in my own words. For the reasons which he gives, the inspector was entitled to go ahead. There is no other ground on which we could disturb his conclusion. Accordingly I agree that the appeal should be allowed.

SIR GEORGE WALLER: I entirely agree with the judgment of Slade LJ. The problems which the inspector had to face permitted of more than one answer. The inspector made a decision and in my view it was one which he was reasonably entitled to make.

Appeal allowed with costs.

R *v* Secretary of State for the Environment, *ex parte* P F Ahern (London) Ltd

QUEEN'S BENCH DIVISION
Roch J
March 22 1989

Enforcement notice — Planning permission — Condition requiring use to cease — Use continuing — Enforcement notice alleged material change of use rather than breach of condition — Inspector decided notice defective — Whether notice capable of correction

In September 1985 planning permission was granted to the applicants' predecessors in title to use premises at 29a Broomfield Street, London E14, as a skip waste transfer station. The permission was subject to a condition that the use be discontinued by August 1987. In September 1987 the applicants' application for planning permission to continue the use was refused by the local planning authority, Tower Hamlets London Borough Council, on the grounds of the effect on local amenities and local policies.

The use continued without the benefit of planning permission and in September 1987 the local planning authority issued an enforcement notice alleging a material change of use without planning permission; in fact the allegation should have been a breach of a condition in a planning permission. Following an appeal by the applicants, the inspector found that the notice was so defective as to be invalid and that he could not exercise the powers of correction in section 88A of the Town and Country Planning Act 1971. Accordingly he allowed the appeal without considering the planning merits.

The applicants moved for judicial review and *certiorari* to quash the decision on the grounds that it was wrong in law. They submitted that they would have suffered no injustice had the substance of the appeal been dealt with, as their evidence and submissions would have been the same. The respondent Secretary of State for the Environment submitted that the error in the enforcement notice was fundamental and outside the power of the Secretary of State to cure; and, that even if this primary submission were wrong, the inspector's decision that he should not amend the notice, because he could not be satisfied, on the balance of probabilities, that no injustice would be caused by amending it, was not unreasonable in the *Wednesbury* sense. The respondent reserved the point, and did not concede that a person in the position of the applicants may proceed by way of judicial review under Order 53 of the Rules of the Supreme Court, rather than by way of an appeal under section 246 of the Town and Country Planning Act 1971.

Held The decision of the Secretary of State for the Environment, by his inspector, was quashed and the matter remitted for further consideration in the light of the judgment.

The early decisions of the courts showed that enforcement notices must be clear as to the alleged breach of planning control: a breach of a condition, or the carrying out of development without planning permission, must be clearly specified. But the introduction of section 88A to the 1971 Act by the Local Government and Planning (Amendment) Act 1981, and the decisions in more recent cases, show that the law has progressed in this field. Under section 88A the Secretary of State can now correct *any* defect "if he is satisfied that the correction or variation can be made without injustice to the appellant or to the local planning authority": see p 105H. Where an enforcement notice alleges a material change of use, rather than a breach of a planning condition, the Secretary of State may exercise his discretion to correct the defect in the notice.

The view of Lord Denning MR in *Miller-Mead* v *Minister of Housing and Local Government* [1963] 2 QB 196, that an error which did not go to the substance of the matter could be corrected so long as the correction could be made without injustice, was now the law: see p 109D. In the present case the inspector should have considered whether a correction could be made without injustice to the parties. The inspector had been wrong in deciding that the error in the enforcement notice was fundamental and incapable of correction; and wrong in holding that he was unable to be satisfied that on the balance of probability no injustice would be caused: see pp 109-110. A B

Cases referred to in the judgment

Associated Provincial Picture Houses Ltd v *Wednesbury Corporation* [1948] 1 KB 223; [1947] 2 All ER 680; (1947) 45 LGR 635, CA

Copeland Borough Council v *Secretary of State for the Environment* (1976) 31 P&CR 403; [1976] EGD 791; 239 EG 503, DC

Coventry Scaffolding Co (London) Ltd v *Parker* [1987] JPL 127, DC C

East Riding County Council v *Park Estate (Bridlington) Ltd* [1957] AC 223; [1956] 3 WLR 312; [1956] 2 All ER 669; (1956) 54 LGR 363; 6 P&CR 231, HL

Eldon Garages Ltd v *Kingston upon Hull Corporation* [1974] 1 WLR 276; [1974] 1 All ER 358; (1973) 72 LGR 456; 27 P&CR 48; [1974] EGD 790; 230 EG 1739

Epping Forest District Council v *Matthews* [1987] JPL 132

Francis v *Yiewsley and West Drayton Urban District Council* [1958] 1 QB 478; [1957] 3 WLR 919; [1957] 3 All ER 529; (1957) 56 LGR 1; 9 P&CR 38; [1958] JPL 154, CA

Kerrier District Council v *Secretary of State for the Environment* (1980) 41 P&CR 284; [1981] JPL 193, DC D

Miller-Mead v *Minister of Housing and Local Government* [1963] 2 QB 196; [1963] 2 WLR 225; [1963] 1 All ER 459; (1963) 61 LGR 152; 14 P&CR 266; [1963] EGD 572; 185 EG 835; [1963] JPL 151, CA

Application for judicial review

This was an application for judicial review by way of *certiorari* to quash a decision of the Secretary of State for the Environment, by his inspector, to allow an appeal against an enforcement notice, a copy of which was served on the applicants. E

Hon Hugh Donovan (instructed by Hamlin Slowe) appeared for the applicants.
Christopher Katkowski (instructed by the Treasury Solicitor) appeared for the respondent, the Secretary of State for the Environment.

The following judgment was delivered F

ROCH J: On May 4 1988 an inspector, appointed by the Secretary of State for the Environment under section 88 of the Town and Country Planning Act 1971 and section 9 of the Local Government and Planning (Amendment) Act 1981, allowed the applicants' appeal against an enforcement notice served on them by the Tower Hamlets London Borough Council. The applicants, far from being pleased at the inspector's decision, take these proceedings to have that decision set aside. The reason for this paradox becomes apparent once the facts of the matter are related. G

The applicants are a commercial undertaking who, as part of their business, operate a skip waste transfer station at 29a Broomfield Street, London E14. These premises are within the London Borough of Tower Hamlets, who are the local planning authority.

Planning permission was granted to the applicants' predecessor in title on September 4 1985. That planning permission was subject to a condition that required the permitted use to be discontinued on or before August 31 1987. The use was such as could give rise to environmental problems, and, for this reason, the local planning authority wished to retain control as to the use of the site. That was the reason given for the condition. H

The applicants were well aware of that condition. Consequently, on March 4 1987, they applied to the local planning authority for planning permission to continue the use of 29a Broomfield Street as a skip waste transfer station. On September 18 1987, the local planning authority refused that application. The reason given for the refusal was:

> The proposal is detrimental to the amenities of local residents by virtue of traffic generation, noise and disturbance and is consequently contrary to policies in the Council's adopted Borough Plan . . .

On September 23 1987 the local planning authority issued an enforcement notice in respect of 29a Broomfield Street alleging that a breach of planning control had taken place and requiring the discontinuance of the use of the premises as a skip waste transfer station and the removal of all equipment and materials used or arising from that use.

The printed form used was that for a material change of use. In fact the printed form which should have been used was that for breach of condition. The Department of the Environment, in their Circular no 38/81, have produced two revised model enforcement notices in the appendices to that circular. One is a model notice where it appears to the local planning authority that there has been operational development without the grant of planning permission required for that development, and the other is a model notice where it appears to the local planning authority that there has been a material change in the use of the land without the grant of planning permission required for that development.

Para 42 of the circular observes:

> . . . these notices are adaptable to deal with other breaches of planning control to which they do not specifically refer.

In fact, Oyez Law Stationers sell enforcement notice forms that are suitable when the breach of planning control is said to be a breach of condition or limitation in a planning permission.

The enforcement notice in this case, in its second recital, said:

> (2) The breach of planning control which appears to have taken place consists in the carrying out of development by the making of the material change in the use of the land described in Schedule 2 below, without the grant of planning permission required for that development.

The first schedule describes the land or premises as:

> 29a Broomfield Street, London E14.

The second schedule reads:

> Alleged breach of planning control (description of the material change of use alleged to have been made)
> use of the said land as a skip waste transfer station

Schedule 3 in the notice reads:

> Steps required to be taken.
> (i) To discontinue using the said land for the purposes of a skip waste transfer station and remove all equipment and materials used or arising from such activity.

In an annex to the enforcement notice, this appears:

> The London Borough of Tower Hamlets consider it expedient to issue this enforcement notice, because the use of the said land as a skip waste transfer station is detrimental to the amenities of local residents by virtue of the noise, disturbance and traffic generation that this use causes.

The words used in that annex were identical to the ground on which planning

permission had been refused on September 18 1987, that is to say five days A
before the issue of the enforcement notice.

The notice was to take effect on October 30 1987, and the applicants were given 28 days from that date to comply with the notice.

The applicants appealed to the Secretary of State against the enforcement notice under section 88(1) of the Act, that appeal being entered on October 15 1987. The grounds of the appeal are set out at p 34 of the agreed bundle. A
subsidiary ground was ground (*h*) in section 88(2), that the period specified in B
the enforcement notice as the period within which any required step is to be taken falls short of what should reasonably be allowed. That ground is not relevant to the present application.

The other ground was ground (*a*) in section 88(2), which reads:

> that planning permission ought to be granted for the development to which
> the [enforcement] notice relates or, as the case may be, that a condition or C
> limitation alleged in the enforcement notice not to have been complied with ought to be discharged.

It is of some significance that ground (*a*) covers four different types of breach of planning control.

The inspector appointed by the Secretary of State held an inquiry into the
appeal on March 1 1988 and inspected the site and the surrounding area on the D
following day. The applicants appeared by counsel who represent them in the present proceedings, Mr Donovan. Four witnesses were called by the applicants and they are set out at p 228 of the agreed bundle.

The local planning authority appeared by counsel, Mr Dayton, and he called a planning officer from the local planning authority's planning department.

The inspector's decision appears at pp 226 and 227 of the agreed bundle of E
documents. What the inspector in fact decided was that the enforcement notice, the subject of the appeal, was so defective as to be invalid and incapable of correction without causing injustice. The notice was therefore quashed and the appeal was allowed. The inspector did not consider the merits or substance of the appeal.

The applicants make these submissions: first, they would have suffered no
injustice had the substance of the appeal been dealt with, as they knew F
precisely what was being alleged against them, namely that they were continuing to use the premises as a skip waste transfer station after August 31 1987, and they knew precisely what steps the local planning authority required them to take.

The applicants say, and I accept it because there is no evidence or any logical
reason on which to arrive at a contrary view, that if the enforcement notice had G
been in the correct form, their evidence and submissions to the inspector would have been precisely the same.

The local planning authority's position is set out in a letter dated August 26 1988, which explains why they do not oppose the applicants' application to this court. That letter appears at pp 230 and 231 of the agreed bundle, and I shall read most of that into this judgment. It is signed by the borough
secretary, and he says: H

> I am able to tell you that I accept the following contentions set out in the Application and you are at liberty to bring this letter to the attention of the Court:
>
> 1. Your Clients' appeal was based upon "ground (*a*)"; that is to say that planning permission should be given to discharge the condition as to limitation of time;
> 2. No issue arose as to the validity of the condition;

> A
> 3. No issue arose as to a material change of use having occurred;
> 4. At the outset of the Inquiry, Counsel for the local planning authority drew attention to the incorrect allegation in the enforcement notice. Counsel for your Clients acknowledged that no injustice would arise from amendment or variation (to allege "breach of condition", rather than "development without planning permission" and he concurred with the application to amend;
> B
> 5. The Inquiry proceeded solely on the issue of whether or not planning permission should be granted for the continuance of the present use (apart from "ground (*h*)" relating to time for compliance).
> It follows that I am satisfied that in the circumstances of this case, the enforcement notice could have been amended without injustice to either party.

As I said at the outset, the applicants are a commercial undertaking. They
C
can see no reason why the substance of this planning matter should not have been decided by the inspector. They wish to know where they stand with regard to these premises. The sooner they have a decision, the better for them as a commercial enterprise. They do not want this matter to be delayed, neither do they wish to have to incur the expense, which they consider unnecessary, of going through the enforcement procedure and an appeal hearing a second time.

D
The Department of the Environment say that it is necessary for the parties to the enforcement notice appeal to go through the procedure again, and for the applicants and the ratepayers of Tower Hamlets to incur the expense of such an exercise because the enforcement notice was defective in a way which they say is incurable. To repeat the words used by the inspector in his decision letter:

E
> This error is in my opinion fundamental and incapable of correction without injustice.

The submission of the respondent's counsel is that the error in the enforcement notice in this case was fundamental. It went to the substance of the matter, and consequently the inspector and the Secretary of State had no power to amend the notice. The error was such that the enforcement notice
F
was invalid in a way which the decided cases show to be beyond remedy. This enforcement notice was like Humpty Dumpty after his great fall — all the king's horses and all the king's men could not put it together again. The question of injustice does not arise, says the respondent. The defect in the notice is one which is altogether outside the power to amend notices conferred on the Secretary of State by section 88A(2) of the Town and Country Planning Act 1971.

G
The respondent makes a second submission in the event that the primary submission is incorrect, namely that if the inspector did have the power to correct the enforcement notice, his decision that he should not do so, because he could not be satisfied on the balance of probabilities that no injustice would be caused by an amendment to the enforcement notice, was not unreasonable in the *Wednesbury* sense; it was neither absurd nor was it perverse.

H
It is necessary to examine the second and subsidiary submission first, because there will be little purpose in examining the law, both statutory and judicial, relating to the respondent's primary submission if the respondent is entitled to succeed on this second submission.

Mr Katkowski, for the respondent, said with great candour that he did not advance the second submission with any great enthusiasm or confidence.

The inspector declined to accept the applicants' assurance at the inquiry that the notice could be corrected without injustice to them because:

> . . . I cannot be certain what actions would have followed had the notice specified the correct allegation in the first instance. Since I cannot be sure of these consequences I am unable to be satisfied that on the balance of probabilities no injustice would be caused.

It is difficult to understand what the inspector had in mind or what actions could have followed had the notice been in the correct form, other than the actions that did follow, namely, an appeal against the enforcement notice on the same grounds on which the applicants did appeal.

Mr Katkowski did try, with considerable courage and ingenuity, to think of circumstances which might have caused some injustice to the local planning authority, but those attempts did not really bear close examination, and Mr Katkowski did not press that part of his argument.

In any event, it is difficult for the respondent to claim that there was any injustice to the local planning authority when it was the local planning authority who were inviting the inspector to amend the enforcement notice and proceed with the hearing of the appeal, albeit the local planning authority were supported in this by the applicants. This was the probable reason why the inspector, in the passage in his decision which I have just quoted, refers only to the applicant and seems to have had in mind a possible injustice to them.

In my judgment, the correction of the enforcement notice would not have caused any injustice to the applicants or to the local planning authority. Both knew the point at issue. The evidence and submissions of both at the hearing of the appeal would have been exactly the same as those the inspector heard. If the matter has to be reheard following the service of a fresh enforcement notice, the evidence and submissions will remain as they were on March 1 1988.

I am satisfied that there was no evidence to support a finding that an injustice might have been caused by the correction of the enforcement notice, and that the inspector's finding, that he was unable to be satisfied on the balance of probability that no injustice would be caused was proven, was, on the facts of this case, perverse and absurd.

Thus it becomes necessary to examine more closely the respondent's primary submission. The submission starts with section 22(1) of the Town and Country Planning Act 1971, which provides:

> . . . "development" . . . means the carrying out of building, engineering, mining or other operations in, on, over or under land, or the making of any material change in the use of any buildings or other land.

The next step is that section 87(1) of the Act makes an apparent breach of planning control after the end of 1963 the first requirement before a local planning authority can issue enforcement notices.

Section 87(3) specifies two categories of breach of planning control:

> (*a*) if development has been carried out . . . without the grant of the planning permission required . . . or
> (*b*) if any conditions or limitations subject to which planning permission was granted have not been complied with.

The development referred to in section 87(3)(*a*) is that defined by section 22(1), that is to say the carrying out of operations or the making of any material change in use.

These two categories are to be seen in section 87(9), which deals with the "steps to be taken in order to remedy the breach" of planning control, because section 87(9) reads:

> . . . steps for the purpose —
> (*a*) of restoring the land to its condition before the development took place . . .

This is a reference to category (*a*) in section 87(3).

Section 87(9) also reads:

> . . . steps for the purpose —
> (*b*) of securing compliance with the conditions or limitations subject to which planning permission was granted . . .

This is a reference to category (*b*) in section 87(3).

The submission is that the two categories are fundamentally different. In category (*a*) there is no planning permission. In category (*b*) there is planning permission, but a condition or limitation of the planning permission has been broken. So far, the respondent's submissions are not controversial.

Then the respondent argues that the foundation for the issuing of an enforcement notice is that it must appear to the local planning authority that there has been a breach of planning control.

Section 87(6) requires:

> An enforcement notice shall specify the matters alleged to constitute a breach of planning control.

Section 87(7) says:

> An enforcement notice shall also specify —
> (*a*) any steps which are required by the authority to be taken in order to remedy the breach;

The respondent says that these are different, depending into which category of the breach of planning control the case falls.

The respondent argues that the decided cases establish, and it is the orthodox view of those who practise planning law, that if the local planning authority specify in an enforcement notice the wrong category of breach of planning control, then that notice is incurably bad and cannot be corrected or varied; it is a nullity or, at the very least, it is invalid beyond redemption.

The reason for this is that the enforcement notice is the start of what can become a penal process. Indeed, initially, an enforcement notice led directly to proceedings before the magistrates. If the local planning authority charge the "wrong offence" in the enforcement notice, the enforcement notice is bad.

At one point in his submission, Mr Katkowski put the matter in these terms: "An enforcement notice is intended to tell the addressee what he has done wrong, and you do not fairly tell a person what he has done wrong if you get the wrong pigeon hole," by which Mr Katkowski meant the wrong category in section 87(3).

I start my consideration of this submission by looking at the statutory development of the enforcement notice procedure. The Town and Country Planning Act 1947, section 75, enabled a local planning authority to serve an enforcement notice for a contravention of previous planning control. The person served with such a notice could complain, by way of appeal to the magistrates under section 28 of the Act, to have the enforcement notice quashed. The magistrates, under section 23(4)(*a*), had to quash the notice if they were satisfied that permission had been granted for the development to which the notice related, or that no such permission was required or, when the notice was based on a breach of condition, that the conditions subject to which permission had been granted had been honoured.

The magistrates could vary the notice under section 23(4)(*b*), if they were not satisfied that the notice should be quashed under section 23(4)(*a*) but they were satisfied that the requirements of the notice exceeded what was necessary to remedy the breach of planning control. In those circumstances they could vary the notice so that only such steps as were necessary to remedy the breach appeared in the notice.

Thus there was no appeal to the Secretary of State or to an inspector A appointed by the Secretary of State, nor was there any statutory power to amend enforcement notices other than the very limited power of variation to which I have just referred.

That was the state of the planning legislation when the House of Lords decided the case of *East Riding County Council* v *Park Estate (Bridlington) Ltd* [1957] AC 223. In that case, Viscount Simonds, at p 233, accepted the general proposition that "the court must insist on strict and rigid adherence to B formalities", because the Act encroached on parties' rights and because disregard of an enforcement notice was an offence involving serious penal consequences. At the bottom of p 234 Viscount Simonds said that because the assertion in the enforcement notice was wrong, the justices were bound to quash the notice. There are similar passages in the speeches of the other members of the House of Lords which decided that if the wrong breach of C planning control was alleged in the notice, then that was fatal to the notice.

The case of *Francis* v *Yiewsley and West Drayton Urban District Council* [1958] 1 QB 478 was also decided under that legislation. Again the court, this time the Court of Appeal, insisted on a strict and rigid adherence to formalities. At p 491 Jenkins LJ said:

> It follows from what I have said that, in my view, the notice proceeded . . . on a wholly false basis of fact; and inasmuch as it proceeded on a wholly false D basis and, so to speak, charged the plaintiff with an offence other than the offence he had actually committed, in my view it cannot stand.

The Caravan Sites and Control of Development Act 1960 came into force on August 29 1960. This Act introduced, for persons served with an enforcement notice under section 23 of the 1947 Act, an appeal to the minister on seven grounds, which are substantially the same grounds as are available today E under the Town and Country Planning Act 1971, as amended. That was done by section 33(1) of the 1960 Act. The section repealed that part of section 23 of the 1947 Act which allowed for appeals to the justices. Most important, for the purposes of this case, are the terms of section 33(5) and (6). Section 33(5) provided:

> On the appeal under this section the Minister may correct any informality, defect or error in the enforcement notice if he is satisfied that the informality, F defect or error is not a material one.

Section 33(6) provided:

> On the determination of an appeal under this section the Minister shall give directions for giving effect to his determination, including, where appropriate directions for quashing the enforcement notice or for varying the terms of the enforcement notice in favour of the appellant.

G

These terms were re-enacted in the Town and Country Planning Act 1971 in section 88(4)(*a*) and (5), which came into force on April 1 1972 and remained in force until November 27 1981, when they were replaced by section 88A(1) and (2), which were substituted for section 88(4)(*a*) and (5) in the 1971 Act by the Local Government and Planning (Amendment) Act 1981, section 1 and Part I of the Schedule.

The leading case on these sections is the Court of Appeal's decision in H *Miller-Mead* v *Minister of Housing and Local Government* [1963] 2 QB 196. In this case Lord Denning MR, at p 221, said:

> Finally, an enforcement notice is no longer to be defeated on technical grounds. The Minister has power under section 33(5) to correct any informality, defect or error in the enforcement notice if he is satisfied that the informality, defect or error is not a material one. This seems to me to be wider than the "slip rule". I think that it gives the Minister a power to amend, which

> is similar to the power of the court to amend an indictment. He can correct errors so long as, having regard to the merits of the case, the correction can be made without injustice. No informality, defect or error is a material one unless it is such as to produce injustice. Applied to misrecitals, it means this: if the misrecital goes to the substance of the matter, then the notice may be quashed. But if the misrecital does not go to the substance of the matter and can be amended without injustice, it should be amended rather than that the notice should be quashed or declared a nullity. In this way the legislature has disposed of the proposition that there must be "a strict and rigid adherence to formalities".

Upjohn LJ, in his judgment at pp 226-227, drew a distinction between an enforcement notice that was a nullity because it was bad on the face of it, for example by being "hopelessly ambiguous and uncertain", and notices that were invalid because, although they appeared to be good, evidence proved that the notices contained errors or defects. In the latter case the notice may be quashed. In the former case, it was unnecessary for Parliament to give the justices power to quash the notice, for it was bad on its face and a nullity.

It is accepted by the respondent that, if Upjohn LJ's distinction is applied to the enforcement notice in this case, it is not bad on the face of it. It is not a nullity. It is invalid.

Upjohn LJ rejected the argument that a misrecital in an enforcement notice *per se* invalidated the notice or that that was the *ratio decidendi* of the *Francis* v *Yiewsley* case. At p 232 Upjohn LJ said:

> I repeat, therefore, that in my judgment the test

— of a valid enforcement notice —

> must be: does the notice tell him fairly what he has done wrong and what he must do to remedy it?

Diplock LJ, in his judgment at p 238, said:

> These two cases,

— he was referring to the *East Riding* case and the *Francis* case —

> which are binding on this court, are in my view authority for the propositions (1) that an enforcement notice, to be valid, must specify whether complaint is made of development without permission required under the Act of 1947, or non-compliance with a condition or limitation subject to which permission under the Act of 1947 was given or, what is now obsolete, non-compliance with previous planning control, and if it fails so to specify, it is a nullity: (2) that an enforcement notice, which alleges development without permission when in fact permission had been given, but a condition or limitation to which the permission was subject has not been complied with, is a bad notice and can be quashed, formerly by the magistrates under section 23(4)(*a*) and now by the Minister under the powers conferred on him by section 33 of the Act of 1960: (3) an enforcement notice which alleges non-compliance with a condition subject to which permission is granted when in fact the development was development without permission, is also a bad notice and can be quashed.

Diplock LJ recognised, at p 239, that the minister has been given wider powers to correct defects and errors in a notice or to vary its terms by the 1960 Act than the justices had under the 1947 Act.

In the *Miller-Mead* case, both enforcement notices contained misrecitals in that they both alleged breaches of planning control greater in extent than the breaches that had actually occurred. The Court of Appeal was unanimous that such misrecitals did not make the notices void and could be corrected by the minister using his then new powers under the 1960 Act.

A second case, decided under identical provisions in the Town and Country Planning Act 1968, is that of *Eldon Garages Ltd* v *Kingston upon Hull Corporation* [1974] 1 WLR 276, a decision of Templeman J, as he then was. This case was not concerned with the minister's powers to amend enforcement notices. The case was concerned with the construction of an enforcement notice. This decision is relied upon by the respondents because of a view stated by Templeman J in several places in his judgment, but finally at p 295B: A

> B My conclusion is that the Act requires the notice to tell the recipient what he is accused of, whether development without permission or breach of a condition, that the magic words used in the Act are not required and the question is whether, on the true construction of the notice, it is clear to the recipient what he has been accused of. I conclude that this notice, on its true construction, does make it clear that it alleges development without permission and *ex abundanti cautela* I conclude that is what the plaintiffs themselves thought it meant. C

The judgment is also instructive because the learned judge in effect adopted the test suggested by Upjohn LJ in the *Miller-Mead* case, namely does the notice tell the recipient fairly what he has done wrong and what he must do to remedy it?

In the final paragraphs of his judgment, at p 295, Templeman J rejected an approach which treated an enforcement notice as a kind of spell by a witch D doctor, and unless the witch doctor got the exact words of the incantation right, then the spell did not work, and said that it was time "that the pettifogging was stopped". In short, the court was rejecting, as had Lord Denning MR in the *Miller-Mead* case, a strict and rigid adherence to formalities.

The 1981 Act inserted section 88A into the Act of 1971. Section 88A(1) reads: E

> On the determination of an appeal under section 88 of this Act, the Secretary of State shall give directions for giving effect to the determination, including, where appropriate, directions for quashing the enforcement notice or for varying its terms.

I read that subsection as relating to the Secretary of State's powers when he has reached a decision. If the evidence has shown that there was no breach of F planning control, he can quash the enforcement notice, or if the appellant has persuaded him that the steps required by the enforcement notice to be taken exceed what is necessary to remedy any breach of planning control or to achieve a purpose specified by the local planning authority under section 87(10) of the 1971 Act or that the period specified in the enforcement notice as the period in which any required step is to be taken falls short of what should reasonably be allowed, the minister can vary the terms of the enforcement G notice accordingly.

Section 88A(2) reads:

> On such an appeal the Secretary of State may correct any informality, defect or error in the enforcement notice, or give directions for varying its terms, if he is satisfied that the correction or variation can be made without injustice to the appellant or to the local planning authority. H

A reading of that subsection, free of authority, would lead me to the conclusion that the Secretary of State had been given the power to correct any informality, defect or error in the enforcement notice, provided that he is satisfied that the correction could be made without injustice to the appellant or to the local planning authority.

Thus, in the present case, the defect or error in the enforcement notice, namely that it recited that the breach of planning control which appeared to the local authority to have taken place consisted in the carrying out of development in the making of a material change of use of the land whereas the breach of planning control which had in fact occurred was a breach of a condition in the original planning permission for this land to be used as a skip waste transfer station, was a defect or error which could be corrected without injustice to either party to the appeal.

The respondent's submission is that section 88A(2) was intended to enact the law as expounded by the courts and not to introduce any change in the law. Thus, says the respondent, it is necessary to look at the old decisions to ascertain the scope of the power to amend given to the minister by section 88A(2).

The respondent goes on to submit that the judgments of Upjohn LJ and Diplock LJ in the *Miller-Mead* case, and the judgment of Templeman J in the *Eldon Garages* case, show that the law before 1981 was that an error or defect in an enforcement notice, stating the wrong breach of planning control, was an error that could not be corrected by the minister using his powers under section 88(5) of the 1971 Act as originally enacted, even if neither party to the appeal would suffer any injustice.

I was referred to many authorities decided since the decision in the *Eldon Garages* case. In some of those cases, such as *Copeland Borough Council* v *Secretary of State for the Environment* (1976) 31 P&CR 403, in the Divisional Court, and *Kerrier District Council* v *Secretary of State for the Environment* (1980) 41 P&CR 284, again in the Divisional Court, there is clear support for the submission being made by the respondent.

For example, in the *Copeland* case, at p 409. Lord Widgery CJ said:

> Unless, however, they get the description of the breach of planning control right in the first instance it seems to me that the enforcement notice is so groggy, if I may use that phrase, right from the outset that it would not be right readily to assume that the planning authority have decided to enforce to a lower standard than was strictly within their powers.

The Lord Chief Justice went on to consider the power of the Secretary of State to correct any informality, defect or error in the enforcement notice if he was satisfied that the informality, defect or error was not material, and concluded that the error in the description of the breach of planning control was gross and said:

> . . . I find it very hard to describe the error in this enforcement notice as being not material.

In the *Kerrier* case, at p 289, Lord Lane CJ said:

> It is clearly established by many decisions that, if the planning authority wishes to serve an enforcement notice, it must decide whether the breach alleged is development without planning permission or failure to comply with some condition or limitation. If, on the facts of any particular case, the planning authority puts the case in the wrong pigeon-hole, the enforcement notice will be set aside.

It is to be noticed that in the *Kerrier* case the court did not have to consider the Secretary of State's power to amend an enforcement notice.

Has the enactment of section 88A(2) in 1981 made any change in the law? There are differences. The first is that the informality, defect or error in the enforcement notice no longer has to be immaterial. Second, the Secretary of State is obliged to consider the question of injustice, not only to the appellant but also to the local planning authority.

Again I have been referred to many decisions made after the enactment of section 88A(2). In some of those decisions the question of whether the word

"variation" is wider than the word "correction" in the subsection has been A
considered, and in others a test has been proposed to determine whether an alteration is a "variation" or not, namely whether there is a sufficient continuity of identity between the notice before and after the change has been proposed and effected.

Two decisions are, in my view, important, and I find them to be of considerable assistance in resolving the question I have to decide in the present
case. The first is *Coventry Scaffolding Company (London) Ltd* v *Parker* B
[1987] JPL 127. This was a decision of the Divisional Court given on June 26 1984: the court consisted of Kerr LJ and Webster J. The local planning authority had issued an enforcement notice alleging unauthorised use of certain premises in Crystal Palace Road, London. The addressees of the notice had more than one premises in Crystal Palace Road, London, but the
enforcement notice omitted to state to which of the premises it related, simply C
describing the premises as Crystal Palace Road, London SE22.

An information was laid before the magistrates charging the company with non-compliance with the notice in contravention of section 89(5) of the Town and Country Planning Act 1971. The magistrates found the offence proved and imposed a modest fine on the owners of the property. The magistrates took the view that the defect in the enforcement notice was not material and
that the company had not suffered any injustice, because all the D
correspondence leading up to the service of the enforcement notice had referred specifically to one of the two properties.

In the course of delivering the judgment of the court, Kerr LJ said at p 131:

> Those authorities showed that Mr Hobson was not correct in his submission that one might only have regard to the contents of the notice, within its four corners, in order to decide whether it was a nullity or not. What had to be
> decided in each instance was whether, in the light of the surrounding E
> circumstances, the recipient of the notice was sufficiently and clearly apprised of the effect of the notice, and of what he had to do pursuant to it to render it just or unjust to hold him to it.

Kerr LJ went on to say that each case would depend on its facts and circumstances.

It was pointed out to me, by Mr Katkowski, that the report of that case does F
not refer in terms to section 88A(2) of the 1971 Act. Whether that is because that section was not brought to the attention of the court, or whether that is because the report is not a full report, cannot be established. What the report does show is that the Divisional Court in that case adopted a pragmatic approach to the question whether or not an enforcement notice was a nullity.

The second authority is the decision in *Epping Forest District Council* v
Matthews [1987] JPL 132. This is a decision of the Divisional Court consisting G
of Stephen Brown LJ, as he then was, and McCowan J. In that case the local planning authority issued an enforcement notice in which para 2 of the recital stated:

> The breach of planning control which appears to have taken place consists in the carrying out of development by the making of the material change in the use of the land described in Schedule 2 below, without the grant of planning
> permission required for that development. H

In the second schedule to the enforcement notice, the alleged breach of planning control was the failure to comply with a condition in the planning consent, namely that the consent should be solely for the benefit of a Mr Gibson. Mr Gibson had died in 1982, and at the time the enforcement notice was issued the caravan was occupied by a Mr and Mrs Matthews. The local planning authority served the enforcement notice on them.

An appeal against the enforcement notice was made by Mr and Mrs Matthews but was dismissed by the Secretary of State, and planning permission to retain a new caravan was refused. Nevertheless Mr and Mrs Matthews continued to occupy the caravan and, on July 30 1985, the justices heard an information brought against Mr and Mrs Matthews. They submitted to the justices that the enforcement notice was bad on its face and was a nullity. The justices accepted that submission and dismissed the case.

The local planning authority appealed to the High Court by way of case stated. It is to be observed that the facts in that case are very close to the facts in the present case.

On the hearing of the appeal to the Secretary of State, the inspector appointed by him had corrected the enforcement notice. It was submitted to the justices, and to the Divisional Court, that the inspector had had no power to correct so fundamental an error and that his purported correction to the notice could not render valid what was a nullity *ab initio*.

Because that notice contained two different allegations of breach of planning control, one arising under section 87(3)(*a*) of the 1971 Act and the other arising under section 87(3)(*b*) of the 1971 Act, it is to be observed that the enforcement notice was bad on the face of it and so was within the first category stated by Upjohn LJ in the *Miller-Mead* case, that is to say a notice which was a nullity as opposed to a notice which was invalid.

Stephen Brown LJ referred in his judgment, with which McCowan J concurred, to section 88A(2) of the Act and is reported, at p 134, as concluding, after referring to the *Coventry Scaffolding* case:

> That case correctly stated the law by which this court was bound, and despite the attractive argument of counsel for the respondents, he [Stephen Brown LJ] was satisfied that what the court before which the validity of an enforcement notice was challenged had to consider was encapsulated in the terms of Upjohn LJ's formulation: "Does the notice tell him fairly what he has done wrong and what he must do to remedy it?" In this particular instance the justices had had before them the enforcement notice. It did tell the recipients quite clearly what they had done wrong and what they had to do to remedy the situation. It was necessary for the justices to read the whole of the notice, including the annex which gave the reasons for it. They should have noted that nowhere in the notice was any factual allegation made against the recipient other than that he had failed to comply with the condition that the "consent shall inure solely for the benefit of the applicant, Mr O F Gibson". The passage in paragraph (2) of the recitals; "consists in the carrying out of development by the making of the material change in use of the land described in Schedule 2 below, without the grant of planning permission required for that development", was plainly on the face of it an erroneous use of terminology. It was also clear that it had never been alleged that the respondents to this appeal had carried out any development. The error apparently arose from the wording left in the pro forma notice used by the clerk who had prepared it. The notice did not include any allegation of fact capable of constituting a material change in the use of the land. Furthermore, paragraph (2) of the recital itself referred to "the material change in the use of the land described in Schedule 2 below". The only matter complained of in Schedule 2 was the failure to comply with the condition in question.
>
> No bench of magistrates could or should have decided that the misuse of language in paragraph (2) of the recitals represented a substantial failure to set out the nature of the allegation made against the respondents. It was clearly an error which could properly be corrected by the Secretary of State using his powers under section 88A(2). In his [Stephen Brown LJ's] judgment the justices were wrongly persuaded that this notice was a nullity by reason of what had been described as the "duality" in paragraph (2) of the recitals and

> Schedule 2. There was not any fundamental ambiguity, and the recipients could not have been under any misapprehension as to what was being complained of and as to what they were required to do to remedy it. Appeal allowed.

This decision shows first that whether or not Mr Katkowski is right, that the view he has urged on me is the orthodox view, the inspector in the *Epping Forest* case was either unaware of or did not subscribe to the orthodox view. Clearly it is not the universal practice for inspectors appointed by the Secretary of State to reach the conclusion that the inspector reached in the present case.

The second thing that this decision shows, in my judgment, is that the law has progressed, both by the enactment of section 88A(2) and recent authority from the law that existed under the former section 87(4)(*a*) and the old decided cases, to the point where the pettifogging has stopped, where artificial and nice distinctions understood only by lawyers no longer prevail, and the Act can be read so that it means what it says, namely that the Secretary of State may correct *any* (which is my emphasis) defect or error in an enforcement notice if he is satisfied that the correction can be made without injustice to either party to the planning appeal.

Of course each case will depend upon its own facts and circumstances, and in those cases where an enforcement notice is defective or in error, and an appellant complains that he has been misled and will therefore suffer injustice, the Secretary of State or one of his inspectors will be slow to correct the defect or error unless satisfied that the notice tells the appellant fairly what he has done wrong and what he has to do to remedy it. If this is the law, as in my judgment it is, then it is no different from the law as stated in 1962 by Lord Denning MR in the *Miller-Mead* case where he said:

> He can correct errors so long as, having regard to the merits of the case, the correction can be made without injustice.

My conclusion is that the inspector in this case was wrong in deciding that the error in the enforcement notice was fundamental and incapable of correction. He was further wrong in holding that he was unable to be satisfied that on the balance of probability no injustice would be caused.

Mr Donovan for the applicants took a further point, namely that the inspector, reaching the conclusions that he did, should either under the procedural rules relating to such appeals, or under the rules of natural justice, have informed the parties of the opinion he was forming, told them of the authorities which he had researched, informed them of his difficulties in concluding that an amendment would cause no injustice to the parties, and have afforded the parties an opportunity to address him further or, at the least, to make written submissions both on the factual issue of whether injustice would result from an amendment and on the further authorities which the inspector was proposing to take into account.

Because of my conclusion on the main issue, I do not need to lengthen this judgment further by considering and deciding that submission. If this matter were to go to appeal, the Court of Appeal would be in as good a position as I am to rule upon that submission.

The final point is one of procedure. The applicants seek their remedy by way of judicial review. The remedy is that of *certiorari* to quash the decision of the inspector and the sending of the matter back to the inspector for his further consideration in the light of this judgment.

It if were material, the respondent would argue that the applicants should have proceeded by way of an appeal under section 246 of the Town and Country Planning Act 1971. However, that would become material only if the applicants were seeking relief in the judicial review proceedings in addition to

A an order for *certiorari*. The applicants commenced their application for judicial review two weeks outside the period for an appeal under section 246 of the 1971 Act. The respondent has indicated, through his counsel, that he does not wish to take any point based on that fact.

As the remedy sought by the applicants is identical to the remedy to which they would be entitled if they had been successful in an appeal under section 246 of the 1971 Act, the respondent has not sought to argue that the applicants
B have brought these proceedings in the wrong form. However, the Secretary of State reserves this point and he does not wish it to be thought that he is conceding that a person in the position of the applicants may proceed either by way of judicial review or by way of appeal under section 246.

Consequently I quash the decision of the inspector and direct that the matter go back to be considered by him in the light of this judgment.

C *Application allowed with costs.*

Strathclyde Regional Council *v* Secretary of State for Scotland

COURT OF SESSION: INNER HOUSE
Lord Ross (Lord Justice Clerk), Lords Wylie and Cowie
April 5 1989

"Person aggrieved" — Structure plan — Alterations — Secretary of State approving alterations subject to modifications — Regional council appealing decision — Whether appeal incompetent — Whether regional council a "person aggrieved" A

In March 1988, the Secretary of State for Scotland approved, subject to modifications, proposals for the alteration of a structure plan prepared by Strathclyde Regional Council. The council appealed the Secretary of State's decision on the grounds that he acted *ultra vires*. On behalf of the Secretary of State there were two pleas in law: (1) the appeal was incompetent in that the appellants were not a "person aggrieved" within the meaning of section 232 of the Town and Country Planning (Scotland) Act 1972; and (2) the crave of the B appeal being incompetent, it should be refused.

Held The first plea in law of the Secretary of State was sustained and the appeal was dismissed.

When section 232 of the 1972 Act was enacted, Parliament had not envisaged that a planning authority would have a right of appeal under that section. Following the decision in *Ealing Corporation* v *Jones* [1959] 1 QB C 384, amendments were made to the legislation to allow local planning authorities to bring appeals in relation to enforcement notices, but no amendments were made in relation to appeals under what is now section 232: see p 115A-C.

There were express rights in sections 233 and 234 of the 1972 Act for an authority to appeal certain actions and orders of the Secretary of State; it is a reasonable inference that Parliament did not intend to confer any such right under section 232: see p 115F. D

It cannot be understood how a planning authority could ever qualify to be a "person aggrieved" in relation to a structure plan, as this is defined as the authority's policy and general proposals and they could hardly be aggrieved by their own statement of policy and general proposals: see pp 116H-117.

Cases referred to in the opinion

Buxton v *Minister of Housing and Local Government* [1961] 1 QB 278; [1960] 3 WLR 866; [1960] 3 All ER 408; (1960) 59 LGR 45; 12 P&CR 77 E
Ealing Corporation v *Jones* [1959] 1 QB 384; [1959] 2 WLR 194; [1959] 1 All ER 286; (1959) 57 LGR 86; 10 P&CR 100; [1959] EGD 178; 173 EG 179
Greater London Council v *Secretary of State for the Environment* [1985] JPL 868
Turner v *Secretary of State for the Environment* (1973) 72 LGR 380; 28 P&CR 123; [1973] EGD 1094; 228 EG 335
Wordie Property Co Ltd v *Secretary of State for Scotland* 1984 SLT 345

Appeal under section 232 of the Town and Country Planning (Scotland) Act 1972 F

This was an appeal under section 232 of the Town and Country Planning (Scotland) Act 1972 against a decision of the Secretary of State for Scotland to approve, with modifications, alterations proposed by the appellants, Strathclyde Regional Council, to their structure plan.

P H Brodie QC (instructed by Simpson & Marwick WS, of Edinburgh) appeared for the appellant local planning authority.

Mrs Ann Paton (instructed by the Solicitor to the Secretary of State for Scotland) appeared for the respondent.

The following opinion of the court was delivered.

LORD JUSTICE CLERK (LORD ROSS): This is an appeal under section 232 of the Town and Country Planning (Scotland) Act 1972 at the instance of Strathclyde Regional Council (hereinafter referred to as "the appellants") against a decision of the Secretary of State for Scotland approving, subject to modifications, proposals for alterations to the structure plan for Strathclyde Region, which decision was communicated to the appellants by letters dated March 10 and 11 1988.

In terms of section 5(1) of the Act of 1972, the appellants, as planning authority, are required to prepare and submit to the Secretary of State for approval a structure plan for their district. Section 5(3) of the Act of 1972 defines the structure plan and provides, *inter alia*, that it shall be a written statement formulating the planning authority's policy and general proposals in respect of the development and other use of the land in the district. Section 7 of the Act of 1972 provides for the approval (in whole or in part and with or without modification or reservations) or rejection of a structure plan by the Secretary of State. Section 8 contains provisions dealing with the alteration of structure plans provides that a planning authority may submit to the Secretary of State proposals for alterations to the structure plan and provides *inter alia* that section 7 shall apply in relation to such proposals as they apply in relation to a structure plan. We were informed that the appellants have had a structure plan since 1981 and that it has been subject to at least two reviews by the Secretary of State.

Mr Brodie for the appellants drew attention to production no 10 in the appeal, which is an update of the written statement of the structure plan dated 1986. As the introduction makes plain, the update written statement deals with a number of matters including, *inter alia*, housing. In the summary, the further proposals in relation to residential development are summarised as follows:

> The Council's comparison of supply and demand has identified an adequate balance within the Greater Glasgow/Lanarkshire Housing Market Area over the period 1986-93. Within the Kyle and Carrick Housing Market Area, there is a small shortfall of one hundred houses.
> Policy RES 13 was added to the Second Review and Alteration to initiate a comprehensive site search with Kyle and Carrick District Council. The policy has been modified to take account of the latest shortfall.

The detailed policies and proposals in this regard are set out in paras 1.52 and 1.53 of the Update (1986) Written Statement.

Mr Brodie also drew attention to production no 11 in the appeal, being the appellants' *Structure Plan Handbook*. He explained that although this was not a statutory document it contained a narrative of all the policies contained in the structure plan as altered or approved from time to time. He drew particular attention to the heading "Strategy" and to the following:

> STRAT 1. THE REGIONAL DEVELOPMENT STRATEGY SHALL BE GENERALLY TO ASSIST EMPLOYMENT BY INCREASING THE ATTRACTION OF THE REGION FOR INVESTMENT, AND IN PARTICULAR TO REDUCE THE FLOW OF EMPLOYMENT AND POPULATION, FROM
> (A) THE CONURBATION BY ACCELERATING THE PROCESS OF URBAN RENEWAL AND REGENERATION
> (B) THE REMOTER RURAL AREAS BY SUSTAINING EXISTING COMMUNITIES AND UTILISING NATURAL RESOURCES.

Mr Brodie also drew attention to pp 26 to 39 in the handbook containing the region's housing policies. It is unnecessary to set these out in detail, but Mr Brodie made two points in this connection. He stressed that the proposals in the Update (1986) Written Statement contained very minor alterations so far as housing policies were concerned. All they did was to provide for an increase in RES 13 of 100 dwellings. In the handbook RES 13 states as follows: A

> LAND WITH CAPACITY TO ALLOW THE DEVELOPMENT OF ELEVEN HUNDRED DWELLINGS BY 1991 WILL BE FOUND IN KYLE AND CARRICK DISTRICT AS FAR AS POSSIBLE ON BROWNFIELD SITES IN ACCORDANCE WITH POLICY RES 1. B

In the Update (1986) Written Statement, RES 13 is stated in the following modified form:

> LAND WITH CAPACITY TO ALLOW THE DEVELOPMENT OF TWELVE HUNDRED DWELLINGS BY 1993 WILL BE FOUND IN KYLE AND CARRICK DISTRICT, AS FAR AS POSSIBLE ON BROWNFIELD SITES IN ACCORDANCE WITH POLICY RES 1. C

The other point which Mr Brodie made was that in neither the handbook nor the Update (1986) Written Statement was there any provision directing that private house building should take place in any particular part of the planning district.

Mr Brodie next drew attention to the provisions of the letters of March 10 and 11 1988. In the former it is stated that the Secretary of State has decided to approve the Strathclyde Structure Plan Update 1986 subject to a number of modifications. So far as the present appeal is concerned he drew attention to the modifications annexed to the letter of March 10 1988 which show that paras 1.52 and 1.53 of the Update (1986) Written Statement are deleted, and in substitution therefor are paras 1.52 to 1.61. Mr Brodie went through these modifications in detail, and in particular he drew attention to the fact that the Secretary of State has made his own assessment of what the supply and demand for owner-occupied housing is. Table 1.8 shows a shortfall of 2,176. This table shows a local surplus in Glasgow, but, after mobile demand is taken into account, there is a mobile shortfall in the area amounting to 2,176. In para 1.61 it is stated, *inter alia*: D E

> The shortfall of 2,200 which the table reveals in the Greater Glasgow/Lanarkshire Area as a whole should be made good in Glasgow, rather than in the rest of the area, in view of the Regional Council's central strategy of regeneration. F

Without going into greater detail, Mr Brodie explained that the appellants' submission put shortly was that by allocating all the additional places to Glasgow under this modification, and by doing this under reference to the appellants' central strategy of regeneration, the Secretary of State had misunderstood the appellants' policy as stated in STRAT 1. STRAT 1 referred to reducing the flow of employment and population from "the conurbation"; it was clear from the glossary of terms on p 122 of the handbook that "the conurbation" meant: G

> The urban areas of the Districts of Dumbarton, Clydebank, Bearsden and Milngavie, Glasgow, Strathkelvin, Monklands, Motherwell, Hamilton, East Kilbride, Eastwood, Renfrew and Inverclyde. H

It was also plain from the glossary of terms that the urban area centred on Glasgow was referred to as "the Inner Conurbation". The Secretary of State, it was said, had erred in proceeding on the basis that the appellants' policy was to reduce the flow of employment and population from Glasgow instead of the wider conurbation as defined. Mr Brodie explained that it was in these circumstances that the appellants had brought this appeal under section 232 of the Act of 1972.

In the appeal the appellants contend that the Secretary of State acted *ultra vires*, that he took into account irrelevant considerations and failed to take into account relevant considerations, and that the modifications which he made were so unreasonable that no reasonable Secretary of State could have made them; they accordingly contended that there was no proper basis to support the modifications and that they should be quashed. In drawing attention to these submissions, Mr Brodie referred to *Wordie Property Co Ltd* v *Secretary of State for Scotland* 1984 SLT 345. Before elaborating his submissions on the merits, however, he recognised that the Secretary of State had stated two pleas in law to the competency of the procedure. These two pleas in law are in the following terms:

> 1. The appeal being incompetent in that the Appellants are not a person aggrieved within the meaning of Section 232 of the said Act, it should be dismissed.
> 2. The crave of the appeal being incompetent, it should be refused.

Mr Brodie under reference to the statutory provisions and to certain English authorities submitted that the appellants were persons aggrieved. Mrs Paton, on the other hand, on behalf of the Secretary of State submitted that the appellants could not be regarded as persons aggrieved within the meaning of the section. Read shortly, section 232(1) provides:

> If any person aggrieved by a structure plan . . . or by any alteration, repeal or replacement of any such plan desires to question the validity of the plan, alteration, repeal or replacement on the ground that it is not within the powers conferred by Part II of this Act or that any requirement of the said Part II or of any regulations made thereunder has not been complied with in relation to the approval or adoption of the plan, alteration, repeal or replacement, he may, . . . make an application to the Court of Session under this section.

In order to determine the question of competency raised by the Secretary of State's first plea in law, it is necessary to consider the provisions of a number of sections in Part XII of the Act of 1972. Section 231(1)(*a*) provides:

> Except as provided by the following provisions of this Part of this Act, the validity of —
> (*a*) a structure plan, a local plan or any alteration, repeal or replacement of any such plan, whether before or after the plan, alteration, repeal or replacement has been approved or adopted; . . . shall not be questioned in any legal proceedings whatsoever.

Counsel were accordingly agreed that if the appellants were to have the right to challenge the decision of the Secretary of State contained in the two letters of March 10 and 11 1988, they would require to show that they were persons aggrieved within the meaning of section 232(1). Mr Brodie recognised that there is no specific provision in section 232(1) enabling a planning authority to make an application to the Court of Session under that section, but he maintained that the words "any person aggrieved" were wide enough to embrace a planning authority. He stressed that a planning authority had a clear interest to challenge any decision of the Secretary of State to amend proposals put forward by them for alteration to the structure plan. He maintained that the appellants were injuriously affected by the Secretary of State's decision and that accordingly they were entitled to bring proceedings under section 232. He founded upon *Ealing Corporation* v *Jones* [1959] 1 QB 384; *Buxton* v *Minister of Housing and Local Government* [1961] 1 QB 278; *Turner* v *Secretary of State for the Environment* (1973) 28 P&CR 123; and *Greater London Council* v *Secretary of State for the Environment and Harrow London Borough Council* [1985] JPL 868.

Mrs Paton, on the other hand, submitted that all these cases other than *Ealing Corporation* v *Jones* were of no assistance. She founded upon the *Ealing Corporation* case and she also relied upon the terms of sections 231, 232, 233 and 234 of the Act of 1972.

In our opinion, having regard to the provisions of the various sections contained in Part XII of the Act of 1972, it is clear that the expression "any person aggrieved" in section 232 does not embrace a planning authority such as the appellants. In our opinion, if Parliament had intended to confer upon a planning authority the right to apply to the Court of Session under section 232, Parliament would have made specific provision to the effect. The language used in section 232(1) falls to be contrasted with the language used in sections 233 and 234. Section 233(1) confers upon any person aggrieved by any order to which the section applies a right to make an application to the Court of Session. However, Parliament has also enacted section 233(2) to deal with the position of an authority which are directly concerned with any order to which the section applies. Section 233(2) is in the following terms:

> Without prejudice to subsection (1) of this section, if the authority directly concerned with any order to which this section applies, or with any action on the part of the Secretary of State to which this section applies, desire to question the validity of that order or action on any of the grounds mentioned in subsection (1) of this section, the authority may, within six weeks from the date on which the order is confirmed or the action is taken, as the case may be, make an application to the Court of Session under this section.

Section 234 deals with appeals to the Court of Session against decisions under section 51. Section 234(1) provides as follows:

> If, in the case of any decision to which this section applies, the person who made the application to which the decision relates, or the planning authority, is dissatisfied with the decision in point of law, that person or the planning authority (as the case may be) may, according as rules of court may provide, either appeal against the decision to the Court of Session or require the Secretary of State to state and sign a case for the opinion of the Court of Session.

In our opinion, since Parliament has expressly provided in sections 233 and 234 for an authority or a planning authority having the right to apply to the Court of Session, it is a reasonable inference that Parliament did not intend to confer any such right upon a planning authority so far as proceedings under section 232 are concerned. If Parliament had intended that a planning authority would have a right to apply to the Court of Session under section 232, in our opinion Parliament would have expressly so provided in the same manner as it did in sections 233 and 234. In the *Ealing Corporation* case, which related to the right of a person aggrieved by an enforcement notice to appeal to a court of summary jurisdiction, Lord Parker CJ stated at p 392:

> Finally, I confess that looking at the statute alone one cannot help feeling that if Parliament had intended the local planning authority to be able to appeal it would have said so clearly.

In that case Donovan J (as he then was) at p 393 stated:

> [Counsel for the respondents says that] Planning authorities have important public duties . . . and . . . that that . . . tends to support the view that they have a right of appeal from the magistrates' decision. I agree that the matter is important, but when this Act was passed in 1947 there were numerous decisions of the court available which show that the expression "person aggrieved" is ambiguous. . . . In those circumstances, one might reasonably think that if Parliament wanted to confer a right of appeal on the local planning authority because this matter is so important, it would have used

> words which placed the matter beyond all doubt as it has done, for example, . . . in sections 61 and 64 of the Income Tax Act 1952, where the position of the subject and the public authority in relation to an appeal from an inferior tribunal is made perfectly clear. I also think there is force in the argument that almost invariably in this Act the local planning authority is called by that name, and I think a better view is that Parliament would have done so in section 23(5) [of the Act of 1947] if it had intended the local planning authority to have a right of appeal from the magistrates.

In our judgment these dicta are entirely consistent with the conclusion at which we have arrived. We have not derived assistance from the other English authorities to which Mr Brodie referred. They contain dicta regarding the question of who may be persons aggrieved within the meaning of certain statutes, and discuss the question of whether it is necessary for such a person to show that some legal right of his has been infringed, and which indicate that "person aggrieved" should not be subjected to any restricted interpretation. In the present case, however, where we are dealing with a planning authority, we are of opinion that these cases are of no real assistance.

In our opinion, Mrs Paton was well-founded in contending that it was clear that, when section 232 was enacted, Parliament had not envisaged that a planning authority would have a right of appeal under that section. The *Ealing Corporation* case was decided in 1959. Thereafter Parliament had the opportunity to make the matter clear if Parliament wished to do so. Mrs Paton pointed out that since 1959 Parliament had changed the position of a local authority in relation to enforcement notices. Enforcement notices had been the matter raised in the *Ealing Corporation* case. So far as the statute was concerned, although there are separate statutes for England and Scotland, the provisions were the same in both countries so far as enforcement notices were concerned. Section 31 of the Town and Country Planning (Scotland) Act 1959 contains provisions dealing with proceedings for challenging the validity of certain orders and decisions. Section 31(2) confers upon an authority directly concerned with any order to which the section applies the right to apply to the Court of Session under that section. Section 31(3) lists the orders to which the section applies and section 34 lists various decisions of the Secretary of State to which the section applies. As originally enacted, no mention is made in the section of enforcement notices, and accordingly they were not covered by the section. By Schedule 9 to the Town and Country Planning (Scotland) Act 1969, section 31(4) of the Act of 1959 was amended by the inclusion of, *inter alia*, a new paragraph (*g*) to the following effect:

> Any decision of the Secretary of State on an appeal to him under section 16(1)(*a*), (*f*) or (*g*) of the Act of 1969 against any enforcement notice.

The effect of that alteration was that Parliament conferred upon an authority a right to make an application to the Court of Session in order to challenge a decision of the Secretary of State on an appeal to him against an enforcement notice. These provisions are now to be found in section 231(3)(*f*) of the Act of 1972.

It is thus plain that when Parliament has wished to extend the rights that an authority has to apply to the Court of Session in order to challenge the validity of certain orders, Parliament has done so expressly, and this reinforces our view that Parliament cannot have intended to confer upon a planning authority a right to apply to the Court of Session under section 232.

There is a further reason for concluding that it cannot have been the intention of Parliament that a planning authority should have a right to apply to the Court of Session under section 232. The opening words of section 232(1) provide: "if any person aggrieved by a structure plan desires to question the

validity of the plan, he may make application to the Court of Session under A
section 232''. As already observed, structure plan is defined in section 5(3) of the Act of 1972 as being a written statement formulating the planning authority's policy and general proposals in respect of the development and other use of land. Since the structure plan contains a statement of the planning authority's policy and general proposals, we do not understand how the planning authority could ever qualify to be a person aggrieved by its own statement of policy and general proposals. This consideration confirms that B
"any person aggrieved" in section 232(1) does not include the planning authority. In these circumstances we are satisfied that the present appeal is incompetent in that the appellants are not a person aggrieved within the meaning of section 232 of the Act of 1972. That is sufficient for the disposal of this appeal. Mrs Paton, for the Secretary of State, also addressed us in support of her second plea in law, which was directed to the competency of the crave of the appeal, but it is not necessary to reach any decision upon that submission. C
Likewise it is unnecessary in the circumstances to express any opinion upon the arguments presented by both counsel to the merits of the appeal. We shall accordingly sustain the first plea in law for the Secretary of State and shall dismiss this appeal as incompetent.

Appeal dismissed as incompetent.